GETTING STARTED WITH THE TI-81 GRAPHING CALCULATOR

NEW ENGLAND INSTITUTE
OF TECHNOLOGY
LEARNING RESOURCES CENTER

GETTING STARTED WITH THE TI-86/85 GRAPHING CALCULATOR

Carl Swenson
Seattle University

John Wiley & Sons, Inc.
New York • Chichester • Weinheim • Brisbane • Singapore • Toronto

Copyright © 1998 by John Wiley & Sons, Inc.

All rights reserved.

Reproduction or translation of any part of this work
beyond that permitted by Sections 107 and 108 of the
1976 United States Copyright Act without the permission
of the copyright owner is unlawful. Requests for
permission or further information should be addressed to
the Permissions Department, John Wiley & Sons, Inc.

ISBN 0-471-25361-8

Printed in the United States of America

10 9 8 7 6 5 4 3 2

Printed and bound by Hamilton Printing Company
Cover printed by The Lehigh Press, Inc.

PREFACE

The purpose of this book is to show how to apply the features of the TI-86 and TI-85 graphing calculators to understand calculus. (If at all possible, use a TI-86; the TI-85 lacks the table feature, and some differential equation capabilities.) The book is divided into five parts, corresponding to common areas of focus in a calculus course. The chapters provide a more specific description of each calculus topic. In general, if you are looking for help on a calculus topic, then use the Table of Contents to find the topic, but if you are looking for help on a calculator command, then start by looking in the Index. Each calculus chapter is intended to be stand alone but they all require an understanding of the basics from Part I Precalculus. Part I is intended as a review; it can be skimmed by experienced users or used as a primer by new users of this calculator.

Most of the examples are taken from the widely used calculus book, *Calculus* by Deborah Hughes-Hallet, et al. I would like to acknowledge and thank the Calculus Consortium based at Harvard and John Wiley & Sons, Inc. for permission to freely use examples from that text.

To the student

Using a graphing calculator can be both frustrating and fun. A healthy attitude when you get frustrated is to step back and say, "Isn't that interesting that it doesn't work." Figuring out how things work can be fun. If you get too frustrated, then it is time to ask a friend or the instructor for help. Make sure you have a phone list of friends with the same calculator.

Part I gives you clear sets of key sequences so that you will become comfortable with how your calculator works. The remaining parts shift into a higher gear and only show you calculator screens as guides for the keystrokes. Your *TI Guidebook* provides a resource if you get stuck; it explains each feature briefly, usually with a key sequence example.

Remember that the *Guidebook* is like a dictionary: there is no story line or context. In this book, the features that you need for calculus will be explained in the context of calculus examples. Other calculator features that are less important to calculus may not be mentioned at all. The mathematical content drives this presentation, not the calculator features.

I have included tips about such things as short-cuts, warnings, and related ideas. I hope you will find them useful.

➤ *Tip: Don't use technology in place of thinking.*

To the instructor

It has been my aim to write a set of parallel form books that span the TI graphing calculator series so that you can allow a variety of these models in your classes. These materials are designed to allow you to focus on the calculus, not the calculator. By having the students use these calculator specific materials, you should be able to greatly reduce the problems of using multiple TI models.

Will these materials take care of all your students? Of course not. There will still be the zealous ones who want the programs in assembly language and the anxious ones who want the buttons pressed for them. These materials are aimed at the middle, giving enough guidance so that most students will be able to work through an example without assistance, but not so specific as to be considered a mindless exercise in pressing keys in the right order.

Programming is not an emphasis of this text. I have included four programs which I feel offer graphical help in understanding concepts. I suggest that you acquire the TI-GRAPH LINK™ and download these programs from the TI archives on the Internet where they will be made available. (See the Appendix for the internet address.) You can distribute them to your class using LINK. The programs have been written with cross-platform compatibility in mind, so that they are almost identical throughout the book series.

➤ *Tip: The TI Volume Purchase Plan can provide you with the TI-GRAPH LINK ™ package and/or an overhead model for classroom use.*

Thanks

I would like to thank Seattle University for its generous sabbatical support during the 1997-1998 academic year which allowed me to prepare these materials, and the members of the Seattle University Mathematics Department with whom I have consulted about their experiences using the TI calculators in various classes. Special thanks go to Janet Mills for taking my place as Chair.

I would also like to acknowledge the debt I owe to Brian Hopkins for his eagle-eyed editing and his cogent contributions to the content.

The Seattle University Information Services Help Desk staff was patient and helpful with my technical computer problems.

The folks at Wiley have been a great help in answering questions and making arrangements. Special thanks to Ruth Baruth who suggested this project; to Sharon Smith, the editor; and to Mary Johenk and Madelyn Lesure for handling details.

Finally, I would like to thank my son Will who provided his computer for editing. The greatest thanks go to my wife Julia Buchholz for her patience and constant support.

Carl Swenson, *swenson@seattleu.edu*

CONTENTS

PART I PRECALCULUS

1. BASIC CALCULATIONS   2

Getting started with basic keys 2
Testing for correct operation 3
A quick tour of the 6x5 scientific keypad 3
Magic tricks to change the keyboard: `2nd` and `ALPHA` 5
Getting around: navigation and editing keys 7
The format of numbers 9
How to put yourself in a good `MODE` 11

2. FUNCTIONS: A FUNDAMENTAL TOOL FOR CALCULUS   13

Formula vs. function notation: using `y(x)=` definitions 13
A summary of menu use 15
Evaluating a function at a point 16
New functions from old 17
Defining families of functions by using lists: `y1={...}x` 18

3. TABLES OF FUNCTION VALUES   19

Lists of function values: `y1({...})` 19
A table of values for a function: `TABLE` 19
A table for multiple functions: table scrolling 20
Find the zero of a function from a table 22
Editing a function formula from inside a table 22
Tables for the TI-85 (optional) 23

4. GRAPHING A FUNCTION — 24

    Basic graphing: `y(x)=`, `WIND` and `GRAPH` 24
    Finding a good window: `ZOOM` 25
    Identifying points on the screen 27
    Reading function values from the graph: `TRACE` 28
    Panning a window 29
    What is a good window? 30
    Can you always find a good window? 31
    Other `ZOOM` options 33
    Graphing inverse functions 34

5. FINDING SPECIAL POINTS ON A GRAPH — 35

    Finding zeros: `ROOT` 35
    Finding extrema: `FMIN` and `FMAX` 36
    Finding the y-axis intercept: `YICPT` 37
    Finding an intersection: `ISECT` 37

6. USING MENU KEYS TO SOLVE AN EQUATION — 39

    Solving a single equation 39
    What if ? analysis of an investment using `SOLVER` 41
    A solver on the home screen: `Solver(` 42
    The no solution message: `ERROR 27 NO SIGN CHNG` 43
    Simultaneous solution of systems of linear equations 44

PART II   DIFFERENTIAL CALCULUS

7. THE LIMIT CONCEPT — 46

    Creating data lists and difference lists 46
    What does the limit mean graphically and numerically? 48
    Speeding ticket: the Math Police let you off with a warning 50

8. FINDING THE DERIVATIVE AT A POINT   51

Slope line as the derivative at a point: `TANLN`   51
The derivative at a point, without the tangent line: `dy/dx`   52
Using a numerical approach to finding the derivative at a point: `der1(...)`   53
Comparing the exact and the numeric derivative   53

9. THE DERIVATIVE AS A FUNCTION   56

Viewing a graph of derivative function   56
The derivative function using the math definition   58
The function that is its own derivative: $y = e^x$ 59
Using lists to estimate a derivative function: `Deltalst(`   60

10. THE SECOND DERIVATIVE: THE DERIVATIVE OF THE DERIVATIVE   61

How to define and graph $f(x)$, $f'(x)$, and $f''(x)$ 61
Looking at the concavity of the logistic curve   62
Creating a numeric second derivative using a table   63

11. THE RULES OF DIFFERENTIATION   64

The Product Rule: $(fg)' = f'g + fg'$ 64
The Quotient Rule: $(f/g)' = (f'g - fg')/g^2$ 65
The Chain Rule   65
The derivative of the tangent function   66

12. OPTIMIZATION   68

The ladder problem   68
Box with lid   70
Using the second derivative to find concavity   72

PART III   INTEGRAL CALCULUS

13. LEFT- AND RIGHT-HAND SUMS — 76

Distance from the sum of the velocity data   76
Using sequences to create a list of function values: `seq(...)`   78
Summing sequences to create left- and right-hand sums   78
Approximating area using the left- and right-hand sums   80

14. THE DEFINITE INTEGRAL — 81

The definite integral from a graph: `∫f(x)`   81
The definite integral as a number: `fnInt(...)`   82
Facts about the definite integral   82
The definite integral as a function: `y=fnInt(...,x)`   84

15. THE FUNDAMENTAL THEOREM OF CALCULUS — 86

Why do we use the Fundamental Theorem?   86
The definite integral as the total change of an antiderivative   86
Viewing the Fundamental Theorem graphically   88
Comparing `nDer(fnInt(...)...)` and `fnInt(nDer(...)...)`   89

16. RIEMANN SUMS — 91

A few words about programs   91
Using the `RSUM` program to find Riemann sums   92
The TI-86 and TI-85 program `RSUM`   93

17. IMPROPER INTEGRALS — 96

An infinite limit of integration   96
The integrand goes infinite   99

18. APPLICATIONS OF THE INTEGRAL — 101

Geometry: arc length   101
Physics: force and pressure   101
Economics: present and future value   103
Modeling: normal distributions   105

PART IV   SERIES

19. TAYLOR SERIES AND SERIES CONVERGENCE   108

The Taylor polynomial program   108
Taylor polynomials for $y = e^x$ 110
Showing the interval of convergence for the Taylor series of the sine function   111
How can we know if a series converges?   112

20. GEOMETRIC SERIES   116

The general formula for a finite geometric series   116
Identifying the parameters of a geometric series   118
Summing an infinite series by the formula   119
Piggy-bank vs. trust   120

21. FOURIER SERIES   121

Periodic function graphs   121
The general formula for the Fourier approximation function   123
A program for the Fourier approximation function   123

PART V   DIFFERENTIAL EQUATIONS

22. DIFFERENTIAL EQUATIONS AND SLOPE FIELDS   128

A word about solving differential equations   128
The discrete learning curve using lists   129
The continuous learning curve $y' = 100 - y$ 131
Slope fields for several differential equations   132
A program for slope fields on the TI-85   134

23. EULER'S METHOD   137

The relationship of a differential equation to a difference equation   137
Euler's method for $y' = -x/y$ starting at (0,1)   138
Euler gets lost going around a corner   140

24. SECOND-ORDER DIFFERENTIAL EQUATIONS — 141

The second-order equation $s'' = -g$ 141
The second-order equation $s'' + \omega^2 s = 0$ 142
The linear second-order equation $y'' + by' + cy = 0$ 143

25. THE LOGISTIC POPULATION MODEL — 146

Entering U.S. population data 1790 - 1940   146
Estimating the relative growth rates: P'/P 146
Using the regression line to rewrite the differential equation   150
The general logistic equation   150
The TI-86 built-in logistic regression feature: `LgstR`   151

26. SYSTEMS OF EQUATIONS AND THE PHASE PLANE — 153

The S-I-R model   153
Predator-prey model   155

APPENDIX AND INDEX

APPENDIX — 159

Complex number form   160
Polar coordinates in the complex plane   160
Parametric graphing   162
Internet address information   164
Linking calculators   164
Linking to a computer   165
Troubleshooting   165

INDEX — 169

PART I
PRECALCULUS

1. BASIC CALCULATIONS
2. FUNCTIONS: A FUNDAMENTAL TOOL FOR CALCULUS
3. TABLES OF FUNCTION VALUES
4. GRAPHING A FUNCTION
5. FINDING SPECIAL POINTS ON A GRAPH
6. USING MENU KEYS TO SOLVE AN EQUATION

1. BASIC CALCULATIONS

The tools used to make numeric calculations have developed from the fingers, to an abacus, to a slide rule, to a scientific calculator. In this chapter we see how to use the TI-86 graphing calculator — a current tool of calculation. If you have used a graphing calculator before, you may only need to skim this chapter. The *TI Guidebook* should also be consulted if you are having difficulty getting started. In this book you will find the references to TI calculator keys and menu choices written in the TI-86 font. The TI-86 font looks like this.

A Note for TI-85 users

If you are a TI-85 user, the keyboard and screens will be similar, but not exactly the same, as those shown in this book. The TI-85 has fewer features than the TI-86, so its menus are shorter. When there is a major difference, it will be pointed out in a parenthetic note or a separate section such as this.

Getting started with basic keys

The ON key

Study the keyboard and press the ON key in the lower left hand corner. You will probably see a blinking rectangular cursor. If you do not, then you may need to set the screen contrast. Even if the cursor is showing, it is a good idea to know how to set the screen contrast: as you use the calculator, the battery will wear down and it will be necessary to adjust the screen contrast. Also, the screen contrast may need to be adjusted for different lighting environments.

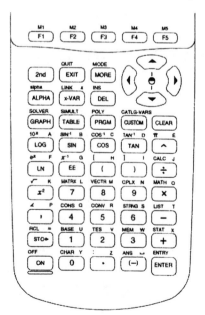

Adjusting the screen contrast

Press and release the yellow key marked 2nd and then press the up arrow gray key in the upper right of the keypad. By repeating this sequence the screen will darken. The screen can be lightened by repeating this sequence but using the down arrow instead of the up arrow. A momentary value (between 0 and 9) will flash in the upper right corner of the screen telling you the battery status (9 is replacement time). If the contrast setting is too low, the cursor will not show, and if it is too high, the screen will be dark as night.

If you take a break and came back later, the cursor will have disappeared for a different reason. The calculator goes to sleep and turns itself off after a few minutes of no activity. Just press the ON key and it will wake up at the same place it turned off: no memory loss. The 2nd_OFF key will turn off the calculator immediately. (This underline notation, 2nd_OFF, means press 2nd then press the key that has OFF written above it.)

➤ *Tip: Sometimes "broken" calculators can be fixed by reinserting the batteries correctly.*

Testing for correct operation

Type some calculation for which you know the answer, say 8*9, and press ENTER; the result appears on the right side of the screen. This graphing calculator has an advantage over a scientific calculator because it shows multiple lines and has entry recall. For example on Figure 1.1, you will see successive entries of three known multiplications. Had they not been all seen together you may not have noticed an interesting pattern — the sum of the answer digits is always 9 (i.e., 7 + 2 = 9, 6 + 3 = 9, and 5 + 4 = 9). On a graphing calculator, your results flow down the screen as you work.

Figure 1.1 First calculations.

➤ *Tip: Test technology with known results before trying complex examples.*

A quick tour of the 6x5 scientific keypad

The keys on the bottom 6 rows (by 5 columns) are the essential keys used on a scientific calculator. Master these first. The gray numeric keys are used for simple arithmetic calculations. Type

$$3 \div 2 \text{ ENTER}$$

Be aware that the symbol on the divide key, ÷, is different from the divide symbol (/) on the screen.

➤ *Tip: Some keypad symbols, like ÷, are displayed on the screen with a different symbol from that on the key.*

Make special note that the gray (-) key is for negation and must be distinguished from the black − subtraction key. One of the most common errors is interchanging the use of the subtraction key − and the negation key (-). Give it a moment's thought: subtraction requires two numbers, while negation works on a single number. Try pressing the following four keys:

3 − 2 ENTER

and then

3 (−) 2 ENTER

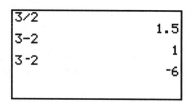

Figure 1.2 Division symbol, and subtraction vs. negation key distinction.

Look carefully at Figure 1.2 which shows these two symbols: subtraction is longer and centered, while negation is shorter and raised. This second entry gives you an answer, but not the one you might have expected: it calculated the multiplication (3)(-2) = -6.

The single most popular error (can errors be popular?) among new users is to not use parentheses when needed. This is a serious error because the calculator does not stop and alert you with an error screen; instead, it gives you the correct answer to a question you are not asking. Suppose you want to add 2 and 6 and then divide by 8. We don't need a calculator to tell us the expression has value 1. But in Figure 1.3 we enter 2 + 6 / 8, and the answer is 2.75. You can figure out that the calculator divided 6 by 8 first and then added that to 2. This was not what we wanted. We need to use parentheses to insure that we are evaluating the correct expression. Now try (2 + 6) / 8 and get an answer of 1 as expected. There is a prescribed order of operations on your calculator; you can look this up in your *TI Guidebook* for details if you have questions about this order.

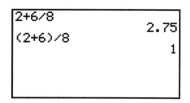

Figure 1.3 Using parentheses to determine the order of operations.

▶ *Tip: When you get an unexpected result, go back and check parentheses — be generous; adding extra parentheses doesn't hurt.*

The black keys: common operations and functions

You probably recognize most of the black operation/function keys. Since exponentiation is usually a raised symbol and word processing is not available on this calculator, the black key ^ is used to signal exponentiation. We know $2^3 = 8$ and to verify this on our calculator, we type

2 ^ 3 ENTER

The LOG and LN are the common log and natural log functions, respectively. The TAN, COS, and SIN keys are the standard trigonometric functions. These functions will appear on the screen in lower case followed by a space but no parenthesis. The parentheses often help clarify an expression; they are found just above the 8 and 9 key. To find the value of log 10, type

LOG 1 0 ENTER

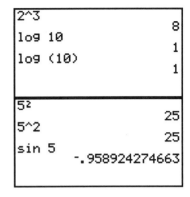

Figure 1.4 Various calculations using the scientific keypad.

▶ *Tip: The parentheses in an expression like log(10) are not actually required, but it is a good habit to include the parentheses in anticipation of cases where the function is part of a more complex expression.*

The effect of the x^2 key is to square a quantity. So pressing

5 x^2 ENTER

will give 25 as an answer. This could also be accomplished by using the general power key ^, specifically 5^2.

The input for trigonometric functions is by default in radian measure. This will be discussed at the end of this chapter.

Magic tricks to change the keyboard: 2nd and ALPHA

How can we do more with the basic keys? There are two tricks: multiple state keys and menus. We first look at different states of the keys. The first change state key is the yellow 2nd key. After pressing it once, the cursor will show an up-arrow on the inside and (presto!) all the keys now have a new meaning. These meanings are indicated in yellow just above and to the left of each key. We have already used the 2nd key to adjust the screen contrast and to turn the calculator off. For a function such as SIN or LOG, the 2nd key gives its inverse.

The second change state key is the blue ALPHA key; as its name indicates, it is used to enter alphabetic letters. We can enter sin 5 without using the SIN key; we can spell it out using the alphabetic keys. Of course this is less convenient in this case. In other situations, especially when using variable names, we will want to type letters directly on the screen.

ALPHA-lock and alpha-lock

The naming of variables is complicated by there being both upper and lower case letters available. Let's start with upper case letters. The ALPHA key, when pressed once, will make only the next key a capital letter. Pressing ALPHA twice in succession will turn on ALPHA-lock: now all successive keystrokes give capital letters and you must press ALPHA a third time (or CLEAR) to return to the normal cursor. The cursor will show an imbedded A in it to indicate that you are in ALPHA mode. Beware that there is no visual change in the cursor between the ALPHA and ALPHA-lock states. Just remember that it is a three-step toggle:

$$\text{Regular} \to \text{ALPHA} \to \text{ALPHA-lock} \to \text{Regular}$$

To enter lower case letters you must first press the 2nd key and then ALPHA. The cursor will show an imbedded a. Pressing ALPHA when in the (lower case) alpha mode results in a lower case alpha-lock. Pressing ALPHA a third time will put you in (capital) ALPHA-lock. Remember that the first change requires 2nd_ALPHA and further changes are made by the ALPHA key. The sequence looks like this:

$$\text{Regular} \to \text{alpha} \to \text{alpha-lock} \to \text{ALPHA-lock} \to \text{Regular}$$

▶ *Tip: The cursor box changes if the ALPHA or 2nd keys are in effect. The 2nd key works as a simple toggle: if you press it by mistake and turn on its state, just press the key again to turn it off. For the ALPHA key, press it twice to turn its state off.*

Stored values STO→

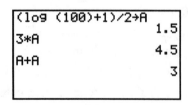

Figure 1.5 Using the store key and a letter variable.

If you wanted to repeatedly use the value of (log(100)+1)/2 in calculations, then you would enter

(LOG (100) + 1) / 2 STO→ A ENTER

as shown in Figure 1.5. The store key, STO→, which appears on the screen with just the arrow symbol (→), is used to save a numeric value into a letter variable.

The variable A can now be used in other computations, as shown in Figure 1.5. You will notice the cursor changes after you press the STO→ key, it is in ALPHA-lock state, in anticipation that you will enter a letter (or word) for a variable. Upper and lower case are distinguished in variable names. Thus A and a are different variables. This is particularly important when we get to functions where x will be the independent variable but X will just be like any other variable name.

The greatest equation ever written: $e^{i\pi} + 1 = 0$

There are five symbols, 0, 1, e, π, i, that we frequently use in mathematics. Incredible as it might seem, they can be related by a single equation. There is no specific key for i, but we know that it stands for $\sqrt{-1}$ so it can be entered that way. The result, (0,0), is an ordered pair which is how complex numbers are shown on the calculator. You can practice using the **2nd** key for e, π, i by entering

```
2nd_e^x ( 2nd_√ ( (-) 1 ) 2nd_π ) + 1
```

Figure 1.6 The equation $e^{\wedge}(i\pi) + 1 = 0$, where (0,0) is seen as zero in complex number format. The value of i is entered two different ways.

In Figure 1.6 a second way to enter this equation is shown where we use (0,1), the ordered pair form of the complex number i.

Getting around: navigation and editing keys

We all make mistakes; correcting them on a graphing calculator is relatively easy. The editing and navigation keys are all located above the 6x5 keypad and below the isolated top row of function keys, **F1, F2, F3, F4, F5**.

Correction keys: DEL, CLEAR, and 2nd_INS

While on the home screen, you can use the gray arrow keys to move forward and backward. (The up arrow takes you to the beginning of the line, the down arrow to the end.) Press the **DEL** key to delete the character in the cursor box. Use **CLEAR** to delete the whole entry line; if the entry line is already clear, then the whole screen will clear. Thus, pressing **CLEAR** twice will always clear the screen. If you need to insert one or more characters, you can move (using the arrow keys) to the location at which you want to insert and press the **2nd_INS** key. This will change the cursor to an underline and signal that the insert mode is in effect. Pressing any arrow key turns off the insert mode.

➤ *Tip: Unlike computer keyboards, these TI calculators have no backspace key.*

Deep recall: 2nd_ENTRY

Often we see errors after we have pressed **ENTER** and left the entry line. The **2nd_ENTRY** key will bring the previous entry line back for us to edit. Even better, by repeatedly pressing the **2nd_ENTRY** key, several of the previous entry lines are accessible. This is called deep recall (the TI-85 has only single recall). Using a previous entry overwrites the current entry; you will not go to a new line. A previous entry cannot be inserted into the command line without erasing what was already there — it is all or nothing.

The pitch of a musical note is determined by the frequency of its vibration (measured in hertz.) Middle C vibrates at 263 hertz. The frequency of a note n octaves above or below middle C is given by $V = 263 * 2^n$. To find the frequency two octaves above middle C and then the frequency two octaves below, we can use the 2nd_ENTRY key to bring back the first calculation and insert a negative sign before the 2 and thus find the frequency two octaves below middle C; see Figure 1.7.

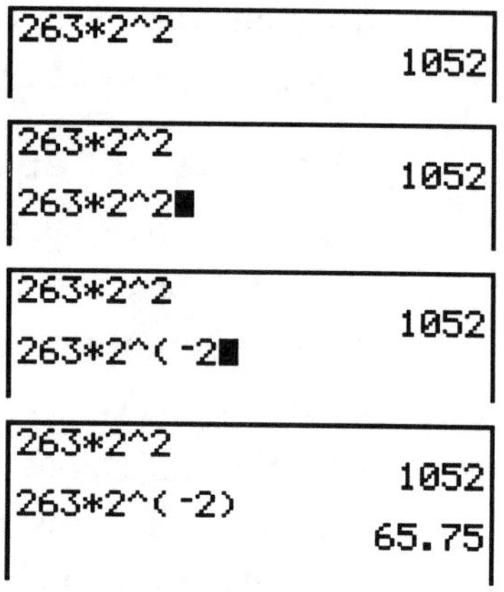

Figure 1.7 Using the recall entry and the insert mode to find two similar calculations. (Partial screens shown.)

Recycling answers: 2nd_ANS

Another keystroke saving feature is the 2nd_ANS key. This places the variable Ans on the screen and uses the previous answer as its value when calculated. To find the difference in area between a 10 and 12 inch pizza, see the first panel of Figure 1.8.

If you start an expression by pressing an operation key, +, -, *, /, ^, the calculator assumes the first number in the calculation is the previous answer, so it puts Ans on the screen without your even pressing the 2nd_ANS key. Try this sequence of keystrokes:

Figure 1.8 Using Ans in the following calculation and using Ans to repeat an action.

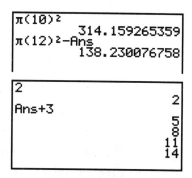

CLEAR 2 ENTER + 3 ENTER ENTER ENTER ENTER ENTER

Using the Ans variable or a value stored in a letter variable insures that you get all the accuracy of the previous calculation. Re-entering a number by hand increases the chance for error; it tempts you to approximate or make a typo.

➤ *Tip: Work smart. Use ENTRY and ANS.*

The format of numbers

Decimals to fractions: ▶Frac

As we proceed we will learn more about menu lines which appear at the bottom of the screen. Here as a first encounter we will follow a set of keystrokes and observe the menu system at work. Suppose that you have entered a calculation and want to convert it to fraction form. Press 2nd_MATH to see a bottom menu line, this assigns meaning to the five keys F1, F2, F3, F4, F5 which are located just below the screen. A decimal calculation and the MATH menu are shown in the first panel of Figure 1.9. Next press MISC (F5) to see a submenu as an additional menu line in the second panel. The right arrow on the bottom line indicates that there are more choices in this submenu. Now press MORE (it is located to the left of the arrow keys) and you will see five new choices. One choice, as shown in the third panel, is ▶Frac (F1) which will paste the expression Ans▶Frac on the screen. Now press ENTER and the decimal number is shown in fraction format. The EXIT key will allow you to back out of menus. The five function keys, along with MORE and EXIT, control the menu system.

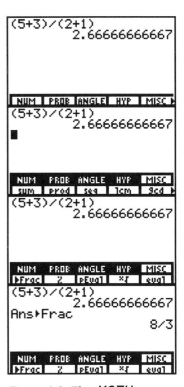

Figure 1.9 The MATH menu used to convert decimals to fractions.

➤ *Tip: The Frac feature will not convert all decimal numbers to fraction form. Conversion is limited to those decimals whose fractions have a denominator of at most three digits.*

If you use really big numbers, they will be displayed in scientific notation. Let's try an unrealistic but surprising situation that uses a big number.

Folding a paper to reach the moon

If you fold a piece of paper, it will double in thickness. You can measure the thickness in sheets: one fold has a thickness of 2 sheets, two folds 4 sheets, three folds 8 sheets, etc. The formula for doubling is $S = 2^n$, where n is the number of folds and S is the number of sheets. We verify the astonishing fact that it takes only 42 folds to reach the moon from earth. (However, it is physically impossible to fold a single sheet more than about seven or eight times.)

Enter 2^42 and find the answer in number of sheets. Had you entered 2^32 or 2^22, the answer would have been given to you in an exact integer form. See Figure 1.10. The distance to the moon measured in sheets of paper exceeds the decimal memory capacity of the calculator, so it gives you the value in scientific notation. We have an answer in trillions:

4.398046511E12 means

$4.398046511 \times 10^{12}$ which is equal to

4,398,046,511,000.

Notice that the exact answer has been rounded to thousands by the fact that we had to add three zeros to the ten significant digits.

Now to verify that this stack reaches the moon, we will first convert sheets to inches by dividing our number of sheets by 300 (an approximate number of sheets per inch) and then by 12 (inches per foot) and by 5280 (feet per mile). This conversion to miles shows that the thickness of the folded paper is more than 230,000 miles, which is the approximate distance from the earth to the moon.

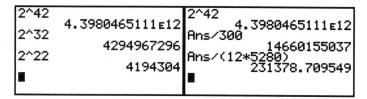

Figure 1.10 Folds to the moon. Large numbers are forced into scientific notation. Finally, conversion of sheets to miles.

Entering scientific notation: EE

If you want to enter five hundred billion without entering all those zeros, you can use the **EE** key to insert the **E** symbol. Press

5 EE 11 ENTER

Note that pressing **EE** results in a single **E** being shown on the screen. This **E** must be distinguished from the variable letter **E** that stores a number. Figure 1.11 first shows that numbers with exponents greater than 11 are put into scientific notation. Next, it shows the variable **E** given the value 2 and that the expressions **5E10** and **5E10** are seen to be different. The latter is an error.

Error screens replace the work screen and tell

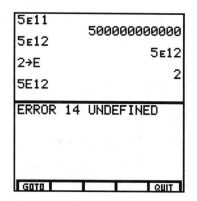

Figure 1.11 Scientific notation has a small **E** *and big* **E** *as a variable will give an error message.*

you briefly what is wrong, and you are forced to respond with GOTO or QUIT. The GOTO option (the F1 key at this point) is usually the best choice because it will 'go to' the error location and allow you to correct it.

How to put yourself in a good MODE

You can control the output format of numeric calculations so that they are all shown in scientific notation. Or, if you are doing a business application, you might want money answers to come out rounded to two decimal places for the dollar and cents format. The 2nd_MODE key allows you to check and change formats. The default settings are all on the left of the screen (see Figure 1.12), so a quick glance will tell you if the any settings have been changed. To change a setting, use the down arrow to reach the desired line and then use the right arrow to move across to the desired setting. You must now press ENTER to make the change. Press EXIT to return to your home screen.

For example, find the total cost of a $1.60 Cafe Latte if the sales tax in Seattle is 8.6%. See Figure 1.12 for the results, where the format has been changed between the first two identical calculations. It should be noted that rounding the result is a display format change only. To see this, multiply the answer by 1000 and you will see that the full decimal accuracy is preserved.

Are your angles in radians or degrees?

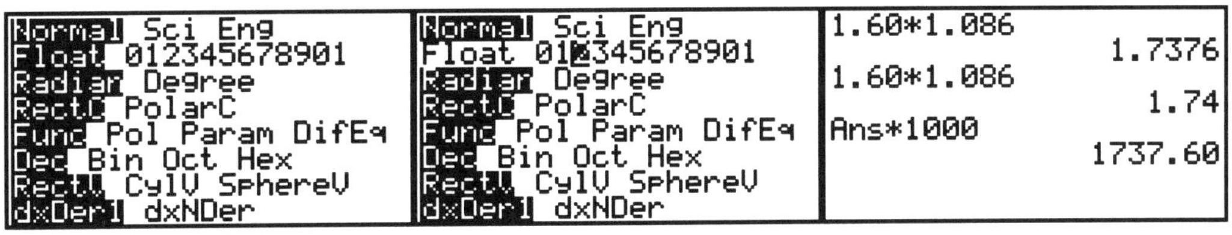

Normal Sci Eng
Float 012345678901
Radian Degree
RectC PolarC
Func Pol Param DifEq
Dec Bin Oct Hex
RectV CylV SphereV
dxDer1 dxNDer
	Normal Sci Eng
	Float 012345678901
	Radian Degree
	RectC PolarC
	Func Pol Param DifEq
	Dec Bin Oct Hex
	RectV CylV SphereV
	dxDer1 dxNDer
		1.60*1.086
		         1.7376
		1.60*1.086
		           1.74
		Ans*1000
		        1737.60

Figure 1.12 Changing the MODE settings to dollar format.

The third row of the MODE screen allows you to switch from Radian to Degree mode for trigonometric function evaluation. Since you are in calculus, you will normally use the Radian setting. For situations where degrees are specified, you can change the setting to Degree. However, it is recommended that you always leave the calculator in Radian mode and use the 2nd_MATH ANGLE menu to paste the degree symbol into the calculation where degrees are used, as shown in Figure 1.13. The last screen shows using the 2nd_MATH ANGLE menu and explicitly setting the angle entry to insure the correct result regardless of the mode setting.

12 PART I / PRECALCULUS

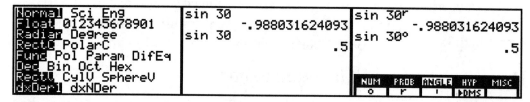

Figure 1.13 The middle screen is deceptive because, between the calculations, the **MODE** *was changed from* **Radian** *to* **Degree**. *The last screen shows how to explicitly specify the angle mode regardless of the mode's angle setting.*

The **MODE** setting choices are described in the *TI-86 Guidebook*, pages 34-36 (pages 1-24 to 1-27 in the *TI-85 Guidebook*). We will mention other settings as we need them, but unless noted otherwise, all our examples will assume that the default settings are in effect.

➤ *Tip: If your output values are in an unexpected or undesirable format, check the* **MODE** *settings. If your* **MODE** *settings do not change, you may have forgotten to press* **ENTER** *before* **EXIT**.

2. FUNCTIONS: A FUNDAMENTAL TOOL FOR CALCULUS

The definition of functions and the use of functional notation is vital to success in calculus. In the next three chapters, we will use the TI calculator to define and evaluate functions, to make tables of values, and to graph functions. In short, we will see how to view functions analytically, numerically and graphically (the Rule of Three). In the previous chapter we used the lower keypad to do our numerical calculations. Now in the next three chapters we will use the top row (just under the screen) of function keys, to activate the GRAPH menu. In this chapter, we'll focus on the first key (F1) in this row which corresponds to y(x)= on the GRAPH menu. It is used to define functions.

Formula vs. function notation: using y(x)= definitions

Function notation is used in calculus, whereas formulas are used in algebra. So what is the difference and how are they related? They both express a relationship between variables. Let's take the famous formula for the area of a circle, $A = \pi r^2$. In precalculus you learned to write this in functional notation $f(r) = \pi r^2$. The functional notation tells you <u>explicitly</u> what variable is the independent variable.

You can define up to ninety-nine functions in your calculator by using the function editor; the editing screen appears when you press the y(x)= key. The available functions are labeled y1, y2, ... , y99. If you enter a function in y1 and press ENTER, then y2 will appear and be ready for a definition. For the moment we ignore the slash (\) that appears to the left of y1; this is a style icon

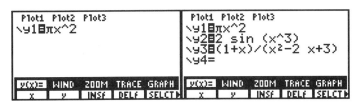

Figure 2.1 Defining functions from the keypad.

which we will use later. To define a function, it is easiest to think of it in formula form. Let one of the y's be the dependent variable and make x the independent variable. For example, in our circle area formula we would define y1=πx^2, where r, the independent variable, is replaced by x.

The independent variable has a special key, x-VAR, that writes the variable x without using the ALPHA key. Functions can be defined using combinations of keys from the 6x5 scientific key area. See examples in Figure 2.1. Notice that implied multiplication, such as 2sin(), will put a space between 2 and sin(). If you return later to see these definitions, the space will have been removed. We see this in the next figure.

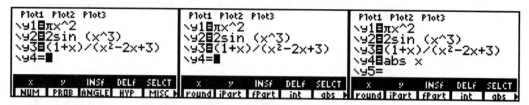

Figure 2.2 Pasting from the 2nd_MATH NUM menu to define abs x.

There are also functions, such as the absolute value, that can be pasted into a function definition from a menu such as the 2nd_MATH NUM menu. Press 2nd_MATH, NUM (F1) and abs (F5) to paste the notation in place at y4=. Finish by pressing x-VAR to add x as shown in Figure 2.2. It should be noted that the menu system has two rows. The bottom row items are selected by pressing the appropriate key F1 to F5. The top row items can be selected by first pressing the 2nd key, as indicated by the yellow M1 to M5 labels above the F keys. We will learn more about the menu system as we go.

▶ Tip: The independent variable must always be called x and this small x is different from the capital X which is a store location label.

Pasting from the CATALOG

Remembering where special symbols and functions are located within various menus can be tedious. This is why there is a built-in alphabetic listing of all functions and settings on the calculator. If you don't know the menus well, then the most convenient way to paste a non-keypad expression into a function is to use the 2nd_CATLG-VAR key. (This key is 2nd_CATALOG on the TI-85 and you will conveniently be in the catalog. There is no need to first press F1 as required for the TI-86.) In the catalog screen you can move quickly to the function you want by pressing the letter key that starts its name. For example, if you want to use the hyperbolic tangent (tanh), then press 2nd_CATLG-VAR CATLG (F1) and you will see the list beginning with abs. Now press T (no need to press ALPHA). Finally use the down arrow to highlight the entry tanh and then press ENTER.

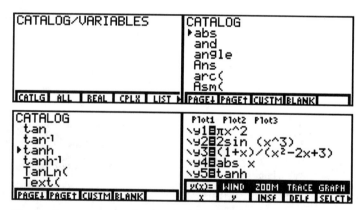

Figure 2.3 Using the 2nd_CATLG-VAR to define function a catalog function.

Cleaning up and getting out: EXIT EXIT or 2nd_QUIT

On the function editor screen, just as for the home screen, use the arrow keys to navigate. Use DEL and INS to edit. Pressing CLEAR will delete any definition. Use EXIT EXIT or 2nd_QUIT to return to the home screen.

➤ *Tip: Be careful about using the CLEAR key. It instantly deletes the entire entry and there is no recovery other than re-entering the formula.*

A summary of menu use

Menus appear on the bottom two rows of the screen as seen in Figures 2.1 to 2.3. We have seen that the GRAPH menu uses a single row, but when we choose y(x)=, the GRAPH menu moves to the top row with the other four options now in black and the bottom row is a submenu related to y(x)=. With two rows showing, you have ten possible choices. The top row can be selected by first pressing the 2nd key.

Figure 2.4 The full GRAPH menu, accessed by the MORE key.

The most important menu by far is GRAPH. You may have noticed that the GRAPH menu has an arrow icon on the far right. This indicates that the menu continues with more items. Pressing MORE will show the next set of five selections or, if you are at the end of the list, then MORE will wrap around and give you the original five menu items. The full GRAPH menu is shown in Figure 2.4.

At the end of the previous section, we used EXIT EXIT or 2nd_QUIT to return to the home screen. The EXIT key first takes you away from the bottom menu row only, so that pressing it once removes the submenu. Pressing EXIT a second time takes you from the GRAPH menu and returns you to the home screen.

➤ *Tip: It is more convenient to use EXIT EXIT instead of 2nd_QUIT.*

In summary, we have seen that menu choices will either paste an item to the cursor location, bring up a new menu line, or bring up a new screen.

The CUSTOM menu

You may have noticed that using the catalog requires several keystrokes. You can store frequently used items more conveniently in your own personal menu, accessed by the CUSTOM key. You are allowed 15 such items. A good thing to put there is y because, as we will see shortly, it is useful to evaluate functions but requires three keystrokes. First give y any value to make it a variable. Then it can

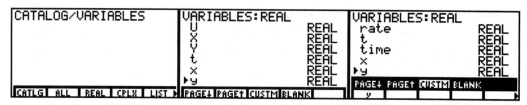

Figure 2.5 **CUSTOM** *menu shown,* **2nd_CATLG-VARS** *items can be placed by using* **CUSTM**. *(Note: store a value in y to make it appear in the* **REAL** *list.)*

be put in the CUSTOM menu as follows: press 2nd_CATLG-VARS, then REAL (F3), move to y in the list, then press CUSTM (F3) and finally press F1 to place it in the position shown in the CUSTOM menu above. See Figure 2.5. The degree symbol (°), buried in 2nd_MATH ANGLE, is another good candidate for your customized menu. It is found by scrolling up above the A listings of the CATLG (F1, not REAL). You may also want to store long variable names that can be pasted as needed. Other helpful items include left and right brackets and double quotes. (The TI-85 only allows CATALOG items to be placed in the CUSTOM menu; you will not be able to store y or any variable names.)

Caution for TI-85 users about evaluating a function

The method of function evaluation given below contains techniques that are not allowed on a TI-85. All calculus students should know that the notation $f(10)$ means evaluate the function f at the value 10; a confused precalculus student might think it means to multiply f by ten. The TI-85 is like the confused student and y1(10) would be evaluated as y1*10. To evaluate y1 at 10 you must use

$$10 \text{ STO} \to x : y1 \quad \text{or} \quad \text{evalF}(y1, x, 10)$$

In other words, the function y1 has a value in terms of the current value of x. The colon in the first command is a convenient way to glue together commands.

Evaluating a function at a point

A benefit of functional notation, $f(x)$, is that $f(3)$ is conveniently understood to be the output value of the function when 3 is input. Each yi stands for the function yi(x), so we can evaluate it at points where it is defined. To find the area of a circle with radius 10 cm, we will use the area function which was our first defined function. (Enter it again if you have deleted it.) To find the value of y1(10) from the home screen, simply enter

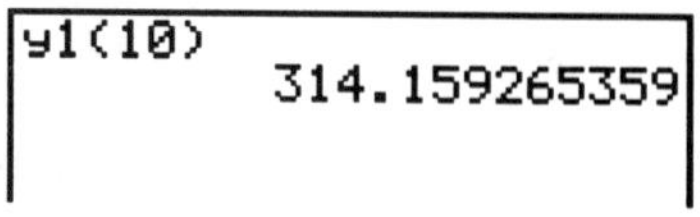

Figure 2.6 Evaluating y1 *at 10 (TI-86 only).*

$$\text{2nd\_ALPHA\_y } 1 \text{ (} 1 \text{ 0) }$$

You can also paste y from the CUSTOM menu if it has been put there.

New functions from old

In the next two examples, we create new functions from previously defined ones.

Composite functions: y1(y2(x))

Suppose an oil spill expands in a perfect circle and that the radius increases as a linear function of time. We can create a new composite function that expresses the area in terms of time. Let $f(r) = \pi r^2$ and $g(t) = 1+t$, where t is in hours. Our new function is $h(t) = f(g(t)) = \pi(1+t)^2$. Using the TI we can create this composite function and find the area of the oil spill after 2 hours. Notice in Figure 2.7 that y3(2) gives the same answer as y1(y2(2)) (using Ans = y2(2)). Even in composite functions all the independent variables must be entered as x.

▶ *TI-85 tip*: *Composition definitions are not allowed on a TI-85. However, individual evaluations can be found by using nested* evalF *or by the following method to find the value shown in Figure 2.7 for* y1(y2(2)): 2 → x : y2 → x : y1

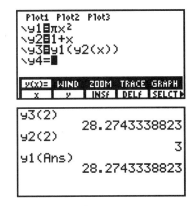

Figure 2.7 Creating a composite function for area in terms of time.

A Malthusian example

In 1798, Thomas Malthus proposed that population growth was exponential and that food supply would grow at a linear rate. We can model a food supply (per million persons) as y1 = 5 + .2x; this means that there is food for five million people in the base year and that each year the supply increases to provide for an additional 200,000 people. (As you enter these functions, the previous ones will be erased.) For the population (in millions), set y2 = 2(1.03)^x; this means that the population starts at two million and increases annually by three percent. Let y3 = y1 − y2 and y4 = y1 / y2. These two new functions are a measure of excess food and a measure of food per capita, respectively. In Figure 2.8 these two measures are evaluated at 50 and 100 years from the base year. There is a shortage in the 100th year (y3 is negative and y4 is less than 1).

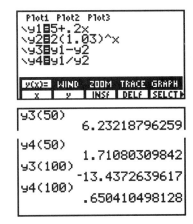

Figure 2.8 Food excess and food per capita in 50 and 100 years.

➤ *Tip: Entering a function in terms of a previous function is easy because* **F2** *pastes in* y.

Defining families of functions by using lists: y1={...}x

Let's see how to define a family of linear functions in just one single function definition. Suppose a city has three taxi companies. Red charges $1.00 to get in one of its taxis and $0.40 for each eighth of a mile traveled. Green charges $2.00 to get in and $0.30 for each eighth of a mile traveled. Blue charges $3.00 to get in and $0.20 for each eighth of a mile traveled. The cost in terms of miles traveled is linear; for example, Red has the cost function $C = 1 + 3.2x$. (For x in whole miles we use $8*0.4 = 3.2$.) We can model this linear situation, $C = b + ax$, with just one function definition whose parameters, a and b, are lists. To enter this example you will need to use the list (or set) symbols { and }; they can be found on the **2nd_LIST** menu. Again, it saves keystrokes to put these in **CUSTOM**.

Notice that the definition does not fit across the screen in Figure 2.9; automatic scrolling leaves y1= but shows you the end of the equation as you type. The left arrow scrolls back to the first part.

Find the charges for one and two mile trips. Notice in Figure 2.9 that the outputs are also lists. We see that Red is cheapest for a 1 mile trip but Blue is cheapest for a 2 mile trip. It should be clear from the rate structure that for longer trips Blue will be cheapest. (TI-85 users will need to evaluate with the method shown earlier.)

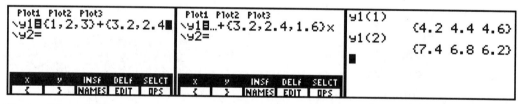

Figure 2.9 *A family of functions created using a list and evaluated simultaneously.*

3. TABLES OF FUNCTION VALUES

This chapter will focus on making lists and tables of function values. These values often reveal the nature of the functional relationship: are the numbers increasing? decreasing? periodic?

Lists of function values: y1({...})

To see a set of values for a function, you can evaluate a function with a list as the input variable The output is a corresponding list of function values. For example, suppose you want to find the area of circles with radii 10 cm, 50 cm, and 100 cm.

- Enter the area function in y1 and type 2nd_QUIT to return to the home screen.
- Evaluate the function with a list as the input: y1({10,50,100}). (Recall that you must enter a lower case y which may be done with 2nd_ALPHA_Y or the CUSTOM menu, as suggested last chapter. Also, the symbols { } are needed to create a list; these can be found under 2nd_LIST or, again, the CUSTOM menu.)
- Use the left/right arrow keys to scroll horizontally when data is too long to fit on the screen. Look for "..." at the right or left of the screen to indicate that there are more values off the screen in that direction.

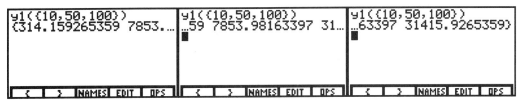

Figure 3.1 With y1=πx², *the values of this area function at 10, 50, and 100. The bottom menu is* 2nd_LIST *to access brackets. Use arrows to scroll the answer set.*

A note to TI-85 users

The following exposition uses tables to show values of a function. The TI-85 does not have this feature. You will need to jump to the last section of this chapter where an alternative method is given to generate lists of function values.

A table of values for a function: TABLE

There is a much more convenient way to see a list of value for a function. We simply enter the function as we just did, set the beginning and increment for a table, and display it. To set up a table, press TABLE and TBLSET (F2). Let's redo the above problem by using a table beginning at zero (TblStart=0) and having the x values go up by ten ($\triangle$Tbl=10). Now press **TABLE** (F1) to display the table. See Figure 3.2. The values beyond 50 are not shown on the first table, but we can use the down arrow to see them.

There is a confusing aspect of this presentation in that 'table' has two meanings. In practice it is not really confusing, but **TABLE** denotes the key and **TABLE** (F1) denotes the first item in the **TABLE** menu.

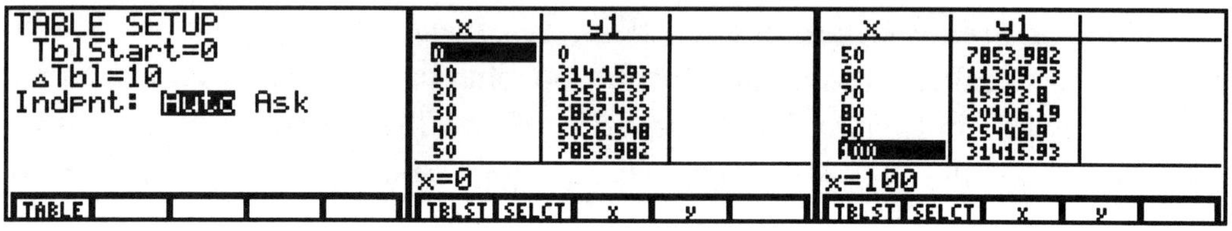

Figure 3.2 Making a table for the area function, y1=πx^2.

Selected values for a table: TBLSET Indpnt: Auto Ask

Suppose we just want the function evaluated at the three specific values 10, 50 and 100. In **TABLE SETUP**, we can arrow down to the option **Indpnt: Auto** and use the right arrow to change from **Auto** to **Ask**. Now press **ENTER** to set the change before you press **TABLE** (F1). The calculator will now ignore the **TblStart** and $\triangle$**Tbl** values and allow you to manually enter the specific values you want to evaluate. See Figure 3.3.

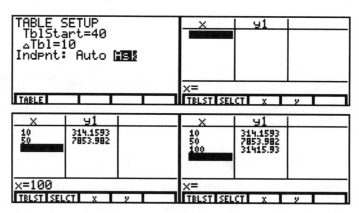

Figure 3.3 How to enter selected values rather than an incremented list.

A table for multiple functions: table scrolling

Recall that in the Malthus models from the previous chapter we had a food supply function and a population function. If we want to enter these again but also want to keep our area function as y1, then we can use y2 and y3 in the y(x)= menu; see Figure 3.4. If you reset the **TABLE SETUP** screen to

TblStart=0 and △Tbl=10 and then press TABLE, you will see the y1 values, which we don't care about right now. Press the right arrow and scroll over to see the y2 and y3 values. Notice that the x values remain on the screen.

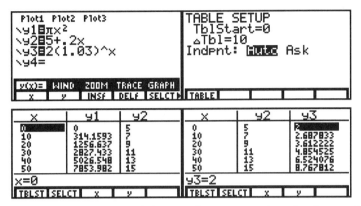

Figure 3.4 Using right/left arrow keys to see selected columns of values.

Selecting and deselecting a function

When you want to see values for only certain functions, you can deselect the ones you do not want and select the ones you do want. On the y(x)= screen, move the cursor to a function definition line and press SELCT (F5). This is a toggle: if it was on (denoted ▪), it will turn off (denoted =); if it was off, it will turn on. When you make a function definition, it is automatically turned on. With y1 deselected, as shown in Figure 3.5, only y2 and y3 show in the table display.

The other four entries that show with SELCT on the y(x)= submenu are x and y, for convenience in defining functions, and INSf and DELf, for inserting and deleting function definitions. For example, you could delete y1 and y2 and the list would then start with y3. To define y1 at this point, you would need to press INSf twice and then both y1 and y2 would show (with no definitions).

It is also possible to deselect a function from the TABLE menu by highlighting the function and pressing ENTER and SELCT (F2). (Notice SELCT appears in both menus but in different F key positions.) But you cannot select a function this way so in general the SELCT key in TABLE is not very useful.

Figure 3.5 Selecting and deselecting a function by pressing SELCT (F5) while the cursor is in the function definition line. Press the TABLE key and TABLE (F1) item.

Find the zero of a function from a table

A natural question arises from the Malthus model: When will the food supply no longer be sufficient for the population? In the previous chapter, we had a function that measured the excess food supply, so we want to find the years for which the excess food function gives negative values. However, it is easier to start by looking for when the excess food function is zero. Enter the relevant functions again and turn off any functions that we don't want to see. Since we know from before that the excess is positive at year 50 and is negative by year 100, we set `TblStart` at 50 and `△Tbl` at 10 (see Figure 3.6). We see that the zero is between year 70 and 80. Now use `TBLST` and set `TblStart` at 70 and `△Tbl` at 1.

You need to arrow down to see that the zero is between 79 and 80. We could continue searching for a more precise value by setting `TblStart=79` and `△Tbl=.1`.

▶ *Tip: Deselected functions are still active for calculations when used in other function definitions.*

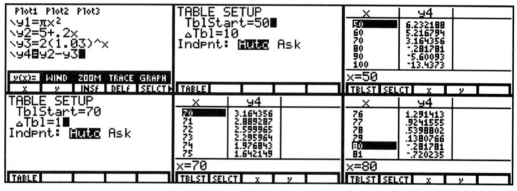

Figure 3.6 Searching for a zero of the food excess function.

Editing a function formula from inside a table

You can redefine a function from the table itself. Suppose you want to look at the food ratio function in the Malthus model. See Figure 3.7. Arrow up to the y4 cell and its equation is shown on the bottom line of the screen. Press **ENTER** to begin editing, arrow right and replace − with ÷, then press **ENTER** to finish editing. The

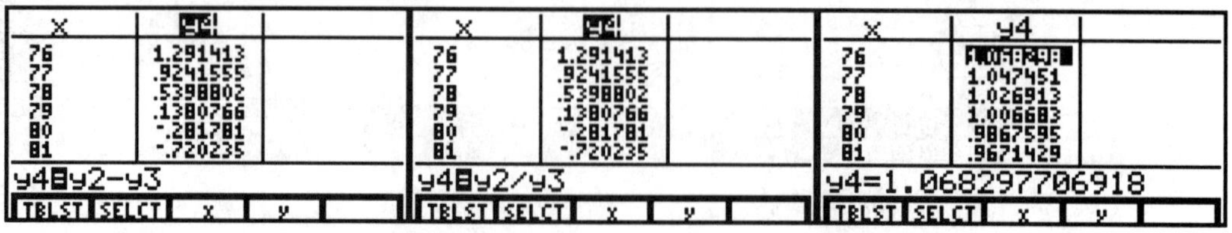

Figure 3.7 Changing a function definition from inside a table.

old y4 formula for excess has now been changed to the new ratio formula, y4=y2/y3. This change has been made permanently as you can verify by looking at the y(x)= definition screen.

Tables for the TI-85 (optional)

This is an optional section for TI-85 users who want to build lists of function values. The lack of the table feature on the TI-85 is one of the main reasons it is quickly becoming obsolete. In the first section of this chapter, you have seen that a function can have a list as input and give a list of function values as output. It is inconvenient to enter a list by hand, so we will start by giving you a command that will generate a list. We will glue a set of commands together using the colon (:). The command set will display ordered-pair values. Then we can use 2nd_ENTRY to modify the command and easily get alternate sets of values. We start with the For(command to repeatedly display (Disp command) the set of ordered pairs (x,y1). The End command marks the end of the For(command. These commands are easily pasted to the home screen from the CATALOG.

Essentially, a For(command repeatedly does the block of commands (here just one) between it and End. The index — x in the example below— takes values from a starting point to a stopping point increasing by an increment. By changing the start, stop and increment, you can control which values of x are shown. See Figure 3.8.

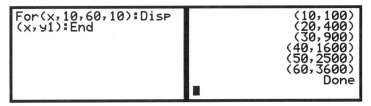

Figure 3.8 With y1=x², *the* For(...) *command produces this list of six function inputs and outputs.*

4. GRAPHING A FUNCTION

This chapter will continue our investigation of the GRAPH menu and show how to graph the functions that we have defined. Like TABLE, discussed in the previous chapter, we have both a GRAPH key, which leads to a menu, and then GRAPH (F5), which draws a graph. The initial five items on the GRAPH menu are fundamental to this chapter:

y(x)=	WIND	ZOOM	TRACE	GRAPH
F1	F2	F3	F4	F5

(The F2 key is called RANGE on the TI-85, but the difference is purely semantic; both allow you to set the range and domain displayed in the graphing window.)

Basic graphing: y(x)=, WIND and GRAPH

Graphing is like the 1-2-3 of taking a picture with a camera.
- *Select your subject(s).* To select a function, recall that you highlight its equal sign on the y(x)= screen (so it appears as ▪). (Deselect or clear functions that you do not want to graph.)
- *Frame them properly.* Press WIND (F2) and set the *x*- and *y*- window boundaries.
- *Click to take the picture.* Press GRAPH (F5).

The hard part of photography is getting the subject both in the picture and looking good. On the calculator we control the picture by using the WIND menu. In Figure 4.1 the first row shows the settings and graph for a picture of the function $y1=\pi x^2$. The function's graph uses too little of the screen. The second row in Figure 4.1 shows an improvement made by changing two settings in the WIND screen.

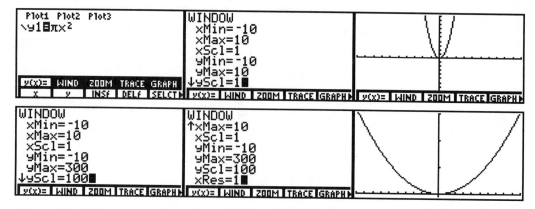

Figure 4.1 The basic graph sequence: y(x)=, WIND, GRAPH *and then an improvement by changing the* WIND *settings. To see the full screen of a graph, press* CLEAR.

You should notice that there is a down arrow to the left of the yScl setting (on the TI-86 only); this signals that there are more items that can be seen by scrolling down. In this case the item called xRes is there; we leave it at the default value of 1. A menu line appears after the graph has been drawn. In the top graph this is not a problem because there is no graph shown in that region. The lower graph has been improved by pressing CLEAR (only once) to remove the menu line (press GRAPH to get the menu).

▶ *Tip: Pressing CLEAR will remove a menu line from the graph screen. Pressing CLEAR CLEAR or EXIT will take you back to the home screen.*

Window settings: xMin, xMax, xScl, yMin, yMax, yScl, xRes

The window setting variables are as follows:
xMin sets the left edge of the window as measured on the horizontal axis,
xMax sets the right edge of the window as measured on the horizontal axis,
xScl (*x*-scale) sets the width between tick marks on the horizontal axis,
yMin sets the bottom edge of the window as measured on the vertical axis,
yMax sets the top edge of the window as measured on the vertical axis,
yScl sets the width between tick marks on the vertical axis, and
xRes (*x*-resolution) sets the selection density of values to plot (1 is the highest setting and should be used unless the graphing is very slow).

Finding a good window: ZOOM

Like photography, the setting of the window is an art. It is rare for us to know the ideal window before we graph; trial-and-error experimentation is usually required. To help with this process, we can use the ZOOM (F3) menu to get started. Use the MORE key to see a new set of five choices. The arrow on the right indicates the continuation of menu choices. Pressing MORE four times brings back the first set of five choices.

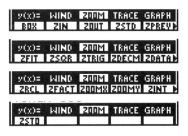

Figure 4.2 The complete ZOOM menu. To see it all requires pressing MORE repeatedly.

With sixteen choices, it can be hard to remember them all. Let's begin with three items that address a common need — a quick window setting.

▶ *Tip: The menus are wrap-around; you can get back to the original set by pressing the MORE key enough times.*

Zoom special settings: ZSTD, ZDECM, ZTRIG

There are three ZOOM menu options that automatically set the window to special settings and graph the selected functions, all in one keystroke. These special settings are shown in Figure 4.3. They are especially helpful when you are graphing common functions (such as polynomial, exponential, and trigonometric) with graphs close to the origin. The ZSTD (zoom standard) often works well as a good first view. The ZDECM (zoom decimal) gives what is called a *nice* window because the *x*-values used to graph progress from -6.3 to 6.3 by tenths. The nicety of this will be explained in the TRACE section just ahead. For trigonometric functions, the obvious first choice for graphing is ZTRIG (Zoom Trigonometry); it uses *x* values from $-(21/8)\pi$ to $(21/8)\pi$ (in decimal form; think of it as -2π to 2π with a little extra on both ends). Using ZSTD will reset xRes to 1, but the other two will not change the xRes setting.

Figure 4.3 ZSTD, ZDECM, *and* ZTRIG *window settings. The* xRes *setting which is not shown has the default setting of 1.*

Window adjustment for Malthus: ZFIT

When graphing functions that model a situation, you will almost always know the domain of the function but probably not the range. If you have entered the domain, you can use ZFIT, a ZOOM option, to help look for the best *y*-values. Let's try this out on the Malthus model of the previous chapters.

   Malthus never published a graph; he used only numerical and analytical expositions. Some of his readers didn't see his concern. A graph appeals to a broader audience. We will now find a good window for the Malthus graphs. Enter or turn on the two Malthus equations. Because we want to look from the base year to a century in the future, we will set xMin=0 and xMax=100. Now the range of the functions is difficult to guess, so we leave the setting in ZSTD (i.e., yMin=-10 and yMax=10). See Figure 4.4.

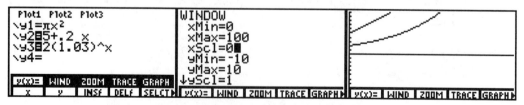

Figure 4.4 Using y(x)=, WIND, *and* GRAPH *to get a first graph with a good domain.*

➤ *Tip: Set* xScl *and* yScl *to zero as you adjust a window, then choose helpful values for them when the final window size is found.*

The window does not show the population growth for all the domain, so we now use ZOOM ZFIT (F1) to reset the *y*-axis window settings so that all the *y*-values of the function will be shown. (The *x*-axis window settings are unchanged by ZFIT.) To see the values that have been set, press

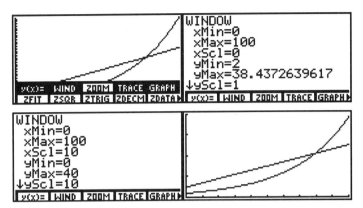

Figure 4.5 Using ZFIT *to find the range, then a reset window with helpful scales.*

WIND. From these values you make a final decision to use yMin=0 and yMax=40. We also set the xScl and yScl at this point to values that will help us read approximate values from the graph. You may need to use CLEAR to see the bottom of the screen. See Figure 4.5.

Since there are no numeric labels on the graph, it is often necessary to alternately use the GRAPH and WIND keys to check on the window dimensions. This problem of knowing where you are on an unmarked graph is the next topic.

Identifying points on the screen

By pressing any arrow key with a graph screen showing, a cross-hair cursor will appear in the center of the screen and the *x*- and *y*-values of the cursor point are displayed. This cursor

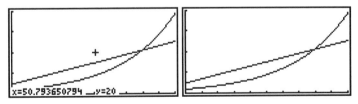

Figure 4.6 The free-moving cursor, activated by an arrow key and deactivated by the CLEAR *key.*

is called the free-moving graph cursor. (Since we are discussing graphs, we will just call it the free-moving cursor.) You can use the arrow keys to navigate this cursor to any point on the screen. By placing the free-moving cursor on the graph of a function, you can display the approximate coordinate values of the point $(x, f(x))$. If you press CLEAR, the free-moving cursor mode is canceled.

Reading function values from the graph: TRACE

You will be more interested in the values of the function than general points on the screen. We improve on the free-moving cursor approach by using the TRACE mode. The TRACE key is F4 in the GRAPH menu and pressing it evokes the trace cursor, a box with a blinking ×. The trace mode displays the x- and y-values of the current trace cursor location. Use the left and right arrows to move to other points on the graph as shown in the second panel of Figure 4.7. The third panel of Figure 4.7 shows the trace cursor being switched to the other graph: the up and down arrows select the function to be traced. The number of the graph being traced is displayed in the upper right corner. In this case we started on y2 and went to y3, for the same x-value. If you press CLEAR, the TRACE mode is canceled.

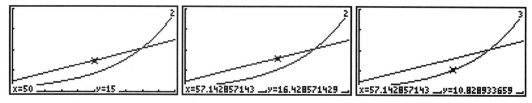

Figure 4.7 The trace cursor on a graph. Right/left arrows move on the graph. Up/down arrows move between selected graphs.

➤ *Tip: Even before displaying the graph screen, the TRACE key will graph and display the trace cursor. So a natural sequence for 1-2-3 graphing is* y(x)=, *WIND, and TRACE.*

Making better trace values: ZINT

Above in the first panel of Figure 4.7 the x-value is 50. As you arrow right using TRACE, you see that the x-values have long decimal expansions. There is little need to know the food supply after exactly 57.142857143 years. If we just wanted a rough idea about values, we could ignore the extra decimal digits. A better option is to use ZOOM ZINT which resets the window so that the trace values will be integers (don't confuse

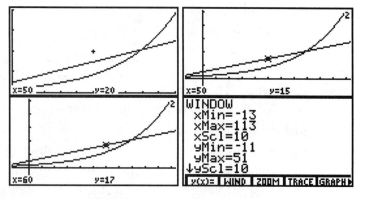

Figure 4.8 ZOOM ZINT sets the x-values to integer values for nice tracing.

this with ZOOM ZIN, discussed below). The ZINT selection is hard to find: use GRAPH ZOOM MORE MORE ZINT (F5). This will display a small cross-hair cursor for you to move to your desired center of the graph. In Figure 4.8, we move it to x=50 and y=20, before pressing ENTER. Use TRACE to see that the *x*-values are now integers. Use WIND to see the new readjusted values.

How to make a nice window (an optional adventure)

The screen is 127 pixels wide and this is why ZDECM is so nice: with xMin=-6.3 and xMax=6.3, the resulting *x*-values are 63 negative tenths, zero, and 63 positive tenths, a total of 127 pixels. Using ZINT in the last example the values of x started at xMin=-13 and ended at xMax=113, this meant a total of 127 by counting zero. For the Malthus graph, if we wanted a nice window that started at xMin=0 then we could set xMax=126. In general, for a graph starting at xMin=0, set xMax=12.6*n, where *n* is large enough to let 12.6*n span the *x*-values you want to include. This produces a nice window!

Panning a window

There is an old story about blindfolded people describing an elephant from different perspectives; their guesses included wall, tree, and snake. Sometimes functions are like elephants: you may need to take the blindfold off to see the whole picture. In case the subject is too big to fit in one window, we can move the window frame to see what is to the left or the right or above or below the current view. This is called panning. Let's take an example of a logistic equation and start as if we knew nothing about it.

$$f(x) = \frac{1000}{1 + 9e^{-0.05x}}.$$

Enter the function as y1=1000/(1+9e^(-.05x)) and follow the twelve steps below to practice panning a window.

1. Enter the function and get a first look by using ZOOM ZSTD.

2. We see no graph. This is the $-10 \leq x \leq 10$ and $-10 \leq y \leq 10$ window.

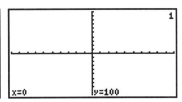

3. Press TRACE to find some point.

(continued next page ...)

30 PART I / PRECALCULUS

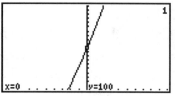

4. Press **ENTER** to 'quick zoom.' This pans the window and places the trace cursor at the center. The quick zoom gives too small a view.

5. Instead, let's try setting the x-values over a broader interval. Set: $0 \leq x \leq 100$, and use **ZOOM ZFIT**.

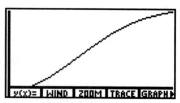

6. The nature of this graph is now more evident. Next we use **TRACE** to explore values.

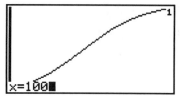

7. To explore around $x = 100$, we press **TRACE 100**. (The TI-85 does not have this feature. Use right arrow.)

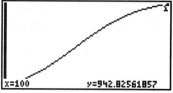

8. This evaluates the function and displays the trace cursor in the upper corner.

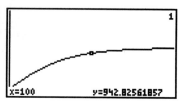

9. Press **ENTER** to pan and move the point $x = 100$ to the center of the graph.

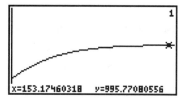

10. The other way to pan is to attempt to trace beyond the right side of the screen; the screen will pan to the right.

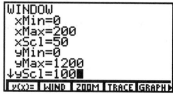

11. Finally, having seen various aspects of the function, press **WIND**, and enter nice integer boundaries.

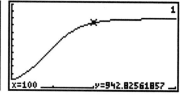

12. Press **TRACE** to see the final view.

So far we have used two techniques to move a window without giving specific numbers. The quick zoom command is **TRACE ENTER**, which pans and centers the window on the trace cursor. Second, the trace cursor, when moved to the left of the **xMin** value or to the right of the **xMax** value, will horizontally pan the window.

➤ *Tip: The free-moving cursor will not pan the screen.*

What is a good window?

You have seen the trial-and-error approach to finding a good window, but this begs the question: what is a good window? For a function serving as a model of a physical situation, a good window will show you the function's graph for the relevant domain. For example, there would be no interest in negative values of the area function $A = f(x) = \pi x^2$. But considering the same function as a purely algebraic quadratic function, we would like to choose a window that includes negative *x*-values so we can see all of its important behavior. Whenever possible, we want to show asymptotic behavior. For example, we needed to see past $x = 100$ in the graph of the logistic function because that function is asymptotic to the line $y = 1000$. We call this the end-behavior. However, if we concentrate solely on the end behavior, we might blur some local behavior. In the logistic example, an important local behavior is that the graph changes from being concave up to being concave down at some point as *x* increases.

Can you always find a good window?

No. There are pathological functions that we cannot graph and others that require move than one view to show all their important behavior. This later case occurs when the end-behavior view makes it impossible to see the local behavior and vice versa.

Asymptotic dangers: Beware of graphs with vertical lines

Sometimes the graphing calculator will lead you astray. The most common case is rational functions. Let's take the blind graph approach to

$$p(x) = \frac{x^2 + 2x + 30}{x - 4}$$

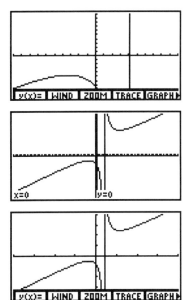

First, we enter y1=(X²+2X+30)/(X-4) (don't forget the parentheses) and use ZOOM ZSTD to get some idea about the function. We need to see a bigger picture.

We now use ZOOM ZOUT. This will display the small cross-hair cursor to designate the center of the zoom out. The current center (the origin) is OK, so press ENTER. The result is a window with the domain and range both four times as big.

Press WIND and improve the crowded ticks on the axes by setting both xScl and yScl to ten.

Figure 4.9 Graphing a rational function.

➤ *Tip: When at the origin, the small cross-hair cursor is hard to see since it appears as a single blinking pixel.*

In Figure 4.9 we are left with an unexplained vertical line to the right of the origin. Could this be part of the graph? A careful look at the function will tell you that the denominator is undefined at $x = 4$. Recall from precalculus that this line is called an asymptote. But the calculator did not draw it as an asymptote. The calculator draws graphs by connecting special x-values that are found by starting at xMin and adding increments of (xMax − xMin) / 127. In this case, to the left of $x = 4$, $f(3.8095238095) = -273.6904762$, and to the right of $x = 4$, $f(4.444444444) = 131.94444446$. Connecting these values gave us the vertical line.

Changing plot style in GRAPH FORMT: DrawLine to DrawDot

This connecting problem can be remedied by changing the plot mode from DrawLine to DrawDot in the third line of the GRAPH FORMT menu as shown in Figure 4.10. Note however that the FORMT item is on the second screen of the GRAPH menu, so the key sequence is GRAPH MORE FORMT. We will later see how to change individual plot styles when needed.

Figure 4.10 Changing the graph format to dot display.

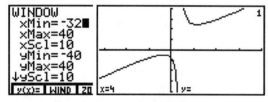

Figure 4.11 Window setting so that the trace cursor shows no y-value at the undefined point $x = 4$. (DrawLine mode.)

A second approach to stop the connection across an undefined point is to create window settings with the undefined value of x exactly in the middle of xMin and xMax. See Figure 4.11. The trace cursor 'shows' that the value is undefined at $x = 4$ by not showing a y-value there.

An inaccurate graph

A rational function graphed on a calculator may connect dots when it should not. A similar distortion occurs when the resolution is insufficient to display a function. Consider

$$f(x) = \sin\left(\frac{1}{x}\right)$$

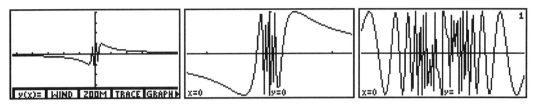

Figure 4.12 Attempts to graph y1 = sin(1/x) *accurately using the three window settings* ZTRIG, ZIN, *and* -.1 ≤ x ≤ .1.

There is no way this function can be accurately graphed if the origin is shown. Several attempts are shown in Figure 4.12.

➤ *Tip: In writing mathematics, it is good style to write decimal values less than one with a leading zero. Therefore, in the caption of Figure 4.12, we should have written -0.1 ≤ x ≤ 0.1. However, when* 0.1 *is entered in the TI calculator, it is converted and shown as* .1. *For this reason, we will often break with style and not use a leading zero.*

Other ZOOM options

For the sake of completeness, we will mention the other choices on the ZOOM menu. The BOX option is similar to ZIN and ZOUT in that it displays the small cross-hair cursor on the graph. Move this cursor with the arrow keys to the screen location where you want a new window to have one corner, press ENTER, then move the cursor to the diagonal corner desired (you will see the rectangle on the screen as you move the cursor), and press ENTER again. The new window will be the rectangle you defined.

The ZPREV option returns you to the previous settings. This is handy if you have changed the window settings and the graph is worse. You can keep one setting in memory: store a window setting with ZSTO and recall it with ZRCL.

The ZSQR selection is helpful for graphing circles. It is similar to ZSTD in that it is a one-keystroke grapher, but the window settings are changed depending upon the current settings: it adjusts the *y* settings so that the *y*-axis has the same scale as the *x*-axis.

The ZDATA entry graphs a good window for statistical data. We will use this only once in a later chapter.

The ZFACT (zoom factor) setting allows you to change the multiplier/divider for ZIN and ZOUT. The default setting of 4 for both axes is good, but a setting of 2 is often convenient as we see in the next section. The ZOOMX and ZOOMY will zoom out only in the indicated direction. Be sure to remember that these zoom out, not in.

Graphing inverse functions

The graph of f^{-1} is the reflection of the graph of f about the line $y = x$. This can be seen in Figure 4.13 for the function $f(x) = x^3$. The window has been set by using ZDECM and then a ZIN after the ZFACT settings are set as 2.

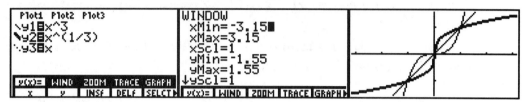

Figure 4.13 The graph of f and f^{-1} as a reflection about the line $y=x$. (The window is ZDECM divided by 2.)

We now graph $f(x) = \sin(x)$ and its inverse, $\sin^{-1}(x)$, in a ZDECM window. We see in the first row of Figure 4.14 that some of the graph's reflection across the $y = x$ line is missing. The difference from the previous example with the cubic function is that the reflection of the sine is not a function. The $\sin^{-1}(x)$ function has to be restricted to make it a function. To draw the full reflection of the function requires a long journey through the menus:

GRAPH MORE DRAW MORE MORE MORE DrInv (F3)

We have turned off the y1=sin x graph so as not to distract our focus on the graph drawn by DrInv. You can see why restrictions are necessary on the domain of $\sin^{-1}(x)$: the reflection is not a function.

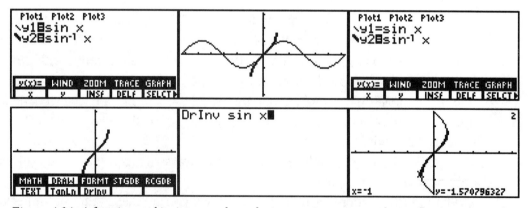

Figure 4.14 A function and its inverse where the inverse is not a complete reflection across the line $y = x$. The DrInv command will draw a complete reflection even thought the graph is not a function. (Uses ZDECM window.)

5. FINDING SPECIAL POINTS ON A GRAPH

We have seen the use of the first five items on the GRAPH menu,

$$y(x)= \quad \text{WIND} \quad \text{ZOOM} \quad \text{TRACE} \quad \text{GRAPH}$$

There is an important item on the GRAPH menu which will be accessible by pressing MORE, it is MATH (F1). In precalculus, the MATH submenu menu helps you locate special points on a graph. Memorize this catchy phrase:

$$\text{GRAPH MORE MATH}$$

In later chapters it will be used frequently to access the calculus items that are on this menu. In this chapter we will restrict our attention to the MATH menu items that are used to identify some special points on a graph.

➤ *Tip: This chapter's references to MATH are all to the GRAPH menu's item MATH (F1). This should not be confused with 2nd_MATH above the multiplication key.*

Finding zeros: ROOT

The first MATH tool, ROOT (F1) finds a zero of a function. We return to the Malthus model (y1 and y2 as shown below) and consider the food excess function y3=y1-y2. The zero of this function has an important meaning: it is when we start having a food shortage.

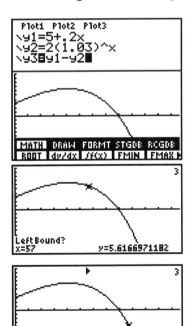

Although not essential, we first turn off y1 and y2 and change the ZSTD window by making xMin=0 and xMax=126 so that more of the excess function values are shown. Then the GRAPH MORE MATH ROOT sequence will start the process of finding the zero. (The TI-85 is slightly different; see the note below this example.)

After selecting ROOT, we are prompted for a Left Bound. Since we are to the left of our zero we just press ENTER. (Optionally, you can enter an *x*-value on the TI-86. The TI-85 only allows the use of the arrows for this setting.)

After pressing ENTER, a right facing arrow on the screen will mark the left bound. Next, you will be prompted to enter a Right Bound. Enter a value of *x* large enough, or arrow to the right far enough, so that the function values are negative there.

(continued next page...)

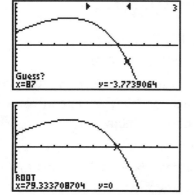

Press ENTER and a left facing arrow will appear to mark the right bound. You are now prompted to Guess, but here you can just press ENTER since there is only one zero in the interval.

The root is shown graphically and given numerically at the bottom of the screen.

Figure 5.1 Zero of a function.

▶ *TI-85 Tip: The Left and Right Bounds are TI-85 MATH menu items LOWER and UPPER and should be set before using ROOT. Set the facing arrows so that the zero is between them. By pressing ROOT, you are given a chance to move the cursor close to your Guess. The Left, Right, and Guess prompts do not have screen labels, so you have to pay special attention to which value you are setting.*

▶ *Tip: The special points of the MATH menu can only be found between the current xMin and Xmax settings.*

▶ *Tip: The closer the bounds and the better the guess, the faster it finds the zero.*

Finding extrema: FMIN and FMAX

The sequence of steps is the same whether you are finding a zero, a maximum, or a minimum. Both FMIN and FMAX use the same sequence as above: Left Bound, Right Bound, Guess.

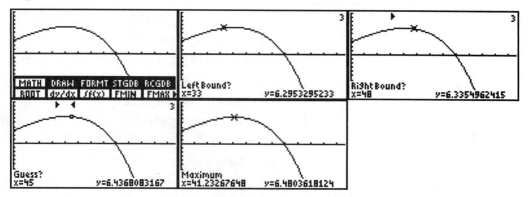

Figure 5.2 Finding the maximum of a function within an interval.

For the excess food function, we might ask in what year the excess is a maximum. By looking at the graph in Figure 5.2 we can see that the maximum is between years 33 and 48. This part of the graph is displayed as a horizontal line segment: this is misleading because the value of the function is not a constant on this interval. Rather, the resolution of the calculator screen is limited. Use the FMAX (F5) selection to find the highest value. We are again asked for the sequence Left Bound, Right Bound, Guess. TI-86 users may enter numeric estimates instead of using the arrows to set the bounds and guess.

➤ *Tip: You can normally press* ENTER *at the Guess prompt, unless you want to speed up the search or you have more than one answer in your interval.*

Finding the *y*-axis intercept: YICPT

To find the *y*-intercept of a function, use YICPT. Remember the key sequence

GRAPH MORE MATH

and add another MORE, then press YICPT (F2). Use the up and down arrows to select a function, if more than one function is graphed, and then press ENTER. The cursor will move to the *y*-intercept point (if that point is in the current window) and the screen will show the intercept value after y=. (It will, of course, also show x=0.)

Finding an intersection: ISECT

A typical task is to find where two functions are equal. Analytically, this means finding the zero of the difference function as we have done using ROOT. Graphically, this means finding where the graphs of the two functions intersect.

We reset the window to $0 \le x \le 100$ and $0 \le y \le 40$ and select y1 and y2 (the food and population functions). Select ISECT from the MATH menu (in the second group of five, so press MORE, then F3). You are asked to identify the first

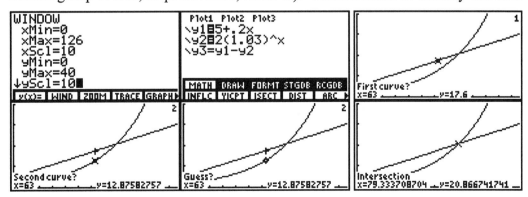

Figure 5.3 Finding the intersection of two graphs.

curve; press **ENTER** to accept the default choice (identified by the flashing trace cursor). Select the second curve by pressing **ENTER**. (This choice is made available in case there are more than two curves graphed. Change a default choice by using the up and down arrows.) Now the calculator prompts you for a guess; press **ENTER** to accept the default. (In cases where there are multiple intersections, you will need to use the left and right arrows to move the cursor to a point near the desired intersection.) We find the intersection point is at 79.333708704 years. We got the same answer as we did using **ROOT** on the difference function — we must have done it right!

➤ *Tip: In cases where you have only two functions graphed and only one intersection appearing in the window, just press **ENTER** three times.*

6. USING MENU KEYS TO SOLVE AN EQUATION

We have just experienced using menus with GRAPH ZOOM and GRAPH MORE MATH. The complete TI menu options are quite extensive and almost as overwhelming as the menu of a gourmet Chinese restaurant. We will not list them here, but anyone wanting to know about the complete set of menu listings can find them in the Appendix of the *TI-86 Guidebook*.

There are three menu keys that help us solve equations. Looking at your keypad you will see the row SOLVER SIMULT POLY in yellow (2nd mode) starting above the GRAPH key.

Solving a single equation

Mathematicians don't use many verbs. We mostly say "this equals that" and shout a few commands like find, evaluate, simplify, and solve. What does it mean to *solve* an equation? For the equation $3x + 2 = 0$, the single solution is $x = -2/3$. The quadratic equation $x^2 - 2x - 3 = 0$ has two solutions, $x = 3$ and $x = -1$. In these examples there is just one variable and we want to know for which value (or values) of the variable the equation is true. If we switch to function notation and let $p(x) = x^2 - 2x - 3$, then solving the quadratic equation is the same as finding the zeros of the function $p(x)$. Thus in this case, and many other cases, the SOLVER is really not essential; you could *solve* the equation by graphing the function and using the techniques of the last chapter to find the zeros. But the GRAPH and ROOT approach can be time consuming, so if you want a quick answer, use SOLVER.

Solving a quadratic equation: SOLVER

Let's find the zeros of the polynomial $p(x)$ given above by using the following five steps.

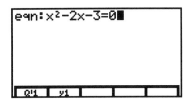

1) Press 2nd_SOLVER and you will see the equation solver screen, where we enter the equation. The bottom menu includes equations that are known, so the list will vary from calculator to calculator. With the equation shown, press ENTER to proceed to the next screen.

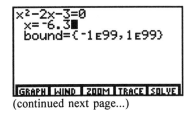

2) In the next stage, you make a guess for x or use whatever the current value of x might be and press SOLVE (F5).

(continued next page...)

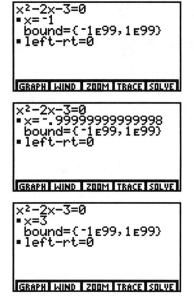

3) The solution is shown with a square mark to indicate it has been calculated. Which of the two solutions you found depends upon your initial guess.

4) To find the other solution, make a guess closer to 3. If we guess x=0, then we get a funny form of the same solution (-1). Some guesses will give close approximations like this and you may need to round to get the exact answer.

5) If you enter a guess of x=2, and press SOLVE the other answer, 3, is shown.

Figure 6.1 Using 2nd_SOLVER.

▶ *Tip: Before you press SOLVE (F5), check to see that the cursor is on the line of the variable for which you want a solution.*

In some cases the calculation will never reach zero exactly; the left-rt indicator (left side minus right side) tells you the exactness of the solution. There are some deep mathematical problems concerning rounding errors and exactness because this is a finite decimal place calculator. In general the results are trustworthy. If you want to restrict the solution search to within some boundaries, you can set bound= with an upper and a lower bound. Setting bound={0,1E99} will insure a positive (or zero) solution.

Solving a polynomial equation using POLY

The 2nd_POLY key offers you the option to choose the order (also called the degree) of the polynomial, fill in the coefficients, and SOLVE for all the solutions at once. This is clearly the method of choice for finding zeros of a polynomial. See Figure 6.2.

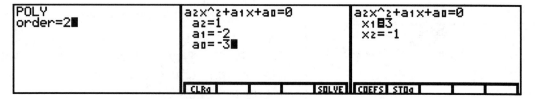

Figure 6.2 Using 2nd_POLY *on the quadratic equation from Figure 6.1.*

What if? analysis of an investment using SOLVER

The previous example of the SOLVER did not show off all of its power. Commonly, equations have several variables, and the SOLVER can be very handy in these cases. For example, the formula for calculating the growth on a continuously compounded investment (a type of exponential growth) is given by

$$P = P_0 e^{kt}$$

where P is the future worth, P_0 is the present value, k is the rate of return, and t is the time (in years) of the investment.

Typically in an investment opportunity, you may be asking any one of four questions:
1. What will my investment be worth at some future date? (Find P, knowing the other variables.)
2. What will I need to investment now in order to get a desired amount in the future? (Find P_0, knowing the other variables.)
3. What investment rate do I need in order to have a desired amount in the future? (Find k, knowing the other variables.)
4. How long will it take my investment to reach a desired amount in the future? (Find t, knowing the other variables.)

Numeric variables on the TI-86 can be from one to eight letters long, so we will choose descriptive names. Let PV stand for the initial (or present value) amount P_0, FV for our future value, rate for our investment rate, and time for time in years.

➤ *Tip: When entering long variable names, use alpha-lock.*

Press **2nd_SOLVER** and enter the new equation as shown below. We will then play "what if" by setting any three of the equation variables and using SOLVE (F5) to calculate the value of the remaining variable.

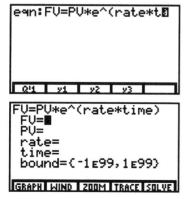

When you enter the SOLVER, any previously used equation will still be there. You may need to use CLEAR to start a fresh equation. (The bottom menu is a known equation list and it will vary.)

When you have entered your equation, press ENTER. A list of your variables and their current values appears — if they are not blank, the variable may have been previously defined.

(continued next page...)

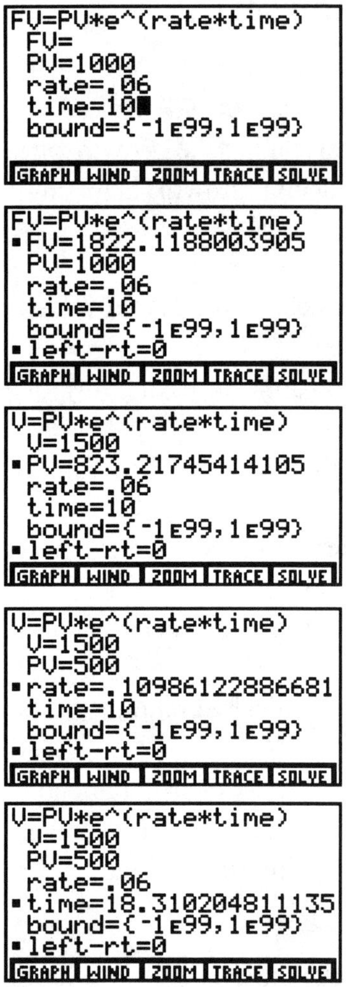

Let's answer a type 1 question: What is the future value of $1000 at 6% in 10 years? To find this, enter the values we know for FV, rate, and time.

Arrow up to FV and press SOLVE (F5). (Answer: $1822.12) If you forget to arrow up to the unknown variable an Error message will appear. Use QUIT and start again with 2nd_SOLVER

To answer a type 2 question, suppose our goal is to have $1500 in 10 years with a 6% rate. Enter 1500 for FV, arrow down to PV, and press SOLVE. You do not need to clear the value of 1000 that exist in PV; it is treated as a guess. (Answer: $823.22)

For a type 3 question, suppose we are promised $1500 in 10 years on an initial investment of $500. What is our rate of return?
(Answer: ≈11%)

Time is the unknown in a type 4 question. This specific answer tells us how long it takes $500 to triple at 6%. (Answer: ≈18.3 years.) This is a general answer in that any amount will take this long to triple.

Figure 6.3 The SOLVER working with an equation of several variables.

A solver on the home screen: Solver(

To solve equations on the home screen, you should first be sure that the command line is cleared and then paste Solver(from the catalog. The general format for the Solver(command is

Solver(*equation, variable, guess*)

The home screen Solver stores the value in the variable but does not print the value on the screen; one technique to see the value is shown in Figure 6.4.

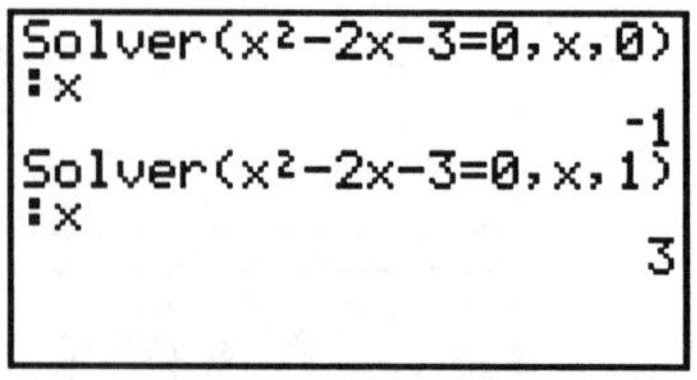

Figure 6.4 Using Solver(to find the zeros of a function from the home screen.

The no solution message: ERROR 27 NO SIGN CHNG

What happens when there is no solution to an equation? For example, find the zeros of the quadratic function $q(x) = x^2 + 2x + 3$ (this is very similar to the polynomial we solved above). Figure 6.5 shows the surprising message ERROR 27 NO SIGN CHNG. This is not really an error, rather the calculator is reporting that it found your expression did not change signs on the interval designated. So what? The way that the SOLVER finds a zero is to find two function values, one positive and one negative. It knows a zero is sandwiched in between and it uses secant lines to hone in on the zero. (If it doesn't start with one positive and one negative value then it still tries to hone in

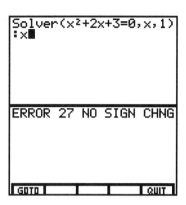

Figure 6.5 The error message when a solution is not found.

on a value within the tolerance of zero.) If it can't find a sign change (or reach a tolerable zero), then it gives you the message about why it failed.

For a quadratic equation, this error message is telling you that there are no real valued solutions. We could also try to use the famous quadratic formula to find the roots. See Figure 6.6. The result is an answer in complex number form, equivalent to $x = -1 + i$.

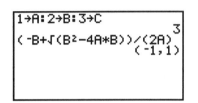

Figure 6.6 Finding a complex number solution from the quadratic equation.

➤ *Tip: Even though the calculator works in complex number mode, the* SOLVER *and* Solver(*will report an error for non-real results.*

A remark about two common errors

Entering the quadratic equation in the last example is like a pop quiz to see if you remember to avoid the two most common entry errors. Be sure to distinguish between negation (-B) and subtraction (B² - 4A*C). Also, use parentheses to place B² - 4A*C entirely inside the square root and, most importantly, use parentheses to demarcate the numerator and denominator. The multiplication 4*A*C requires that there be a multiplication sign between the variables A and C. If it is not there, the calculator assumes there is one variable named AC. The multiplication sign between the 4 and A is optional since it knows to multiply any number in front of a variable. Always enter a fraction as (...)/(...) unless the numerator or denominator is a single number or variable.

Simultaneous solution of systems of linear equations

The third solver key is **SIMULT**. It is used when there is a set of n linear equations with n unknowns. The need to solve such a set of equations is quite common. As an example, differential equations often have general solutions in which constants are unknown, and certain known conditions allow us to solve for these constants.

Suppose we know that a general solution is

$$s(t) = C_1 e^{-t} + C_2 e^{-2t}$$ with related equation $$s'(t) = -C_1 e^{-t} - 2C_2 e^{-2t}$$

and we also know $s(0) = -0.5$ and $s'(0) = 3$. To find our constants C_1 and C_2 we can substitute the specific function values into the general equation and have a system of two equations and two unknowns. You can verify that we get:

$$C_1 + C_2 = -0.5$$
$$-C_1 - 2C_2 = 3$$

In this case you could just use simple algebra to find the solutions: $C_1 = 2$ and $C_2 = -2.5$. But we will see how the TI will find this solution, and then be ready for systems of larger order. The TI can solve systems of order up to 30.

By pressing **2nd_SIMULT** you will be greeted with the first screen shown in Figure 6.7, where the number of equations (or unknowns) is given. Next it will prompt you to enter the coefficients and constant term of each equation. Use **NEXT (F1)** and **PREV (F2)** to scroll through the equations if desired. Once all the equations are entered, press **SOLVE (F5)** and the solution will be displayed.

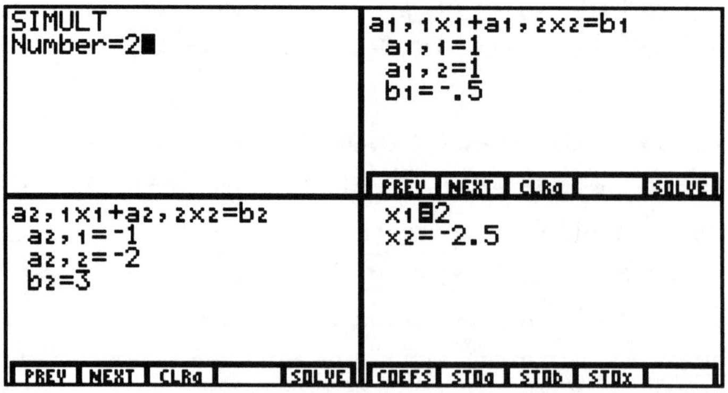

Figure 6.7 Using SIMULT to find the simultaneous solution of a system of 2 equations and 2 unknowns.

PART II
DIFFERENTIAL CALCULUS

7. THE LIMIT CONCEPT

8. FINDING THE DERIVATIVE AT A POINT

9. THE DERIVATIVE AS A FUNCTION

10. THE SECOND DERIVATIVE:
    THE DERIVATIVE OF THE DERIVATIVE

11. THE RULES OF DIFFERENTIATION

12. OPTIMIZATION

7. THE LIMIT CONCEPT

A fundamental difference between precalculus and calculus is the application of the limit. In precalculus, we can define average velocity over a time period of some positive length. For example, if you drive 200 miles in four hours, then you averaged 200/4 = 50 miles per hour. But looking at the car's speedometer, you see the speed at a given moment, the instantaneous velocity.

The following table shows heights of a grapefruit thrown in the air. Find the average velocity over periods of one second. (This is quite simple, but it allows us to cover several general aspects of lists on the calculator.)

time (seconds)	0	1	2	3	4	5	6
height (feet)	6	90	142	162	150	106	30

Figure 7.1 Grapefruit height per second.

Creating data lists and difference lists

Creating lists by hand: {...}

First we put the data in lists using the set bracket notation. This is shown in Figure 7.2, where a **CUSTOM** menu is used to access the set brackets (otherwise they are in **2nd_LIST**). We store the two lists into the variable names **LT** and **LH**. It is a good idea to begin list names with the letter **L** so that they are easily identified. You may also want to call them something more descriptive like **LTIME** and **LHEIGHT**.

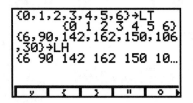

Figure 7.2 Entering the data into two lists by hand.

➤ *Tip: When you press the* **STO→** *key, the calculator is in* **ALPHA**-*lock mode so that the letter* **L** *is a single keystroke. If you want to enter a number, you must first press* **ALPHA** *to deactivate* **ALPHA** *mode. In general it is easier to enter list names that are all letters.*

7. THE LIMIT CONCEPT   47

Creating lists as columns: 2nd_STAT EDIT

The preferred method of entering lists is 2nd_STAT EDIT. This allows entry in parallel columns and the ease of scrolling. When you first see this screen, the column headings may be the default lists, xStat, yStat, fStat, but you can arrow to the up to the heading names and arrow right to an empty heading. Enter the list name and press the down arrow to begin entering data. If you have already entered data in a list, then

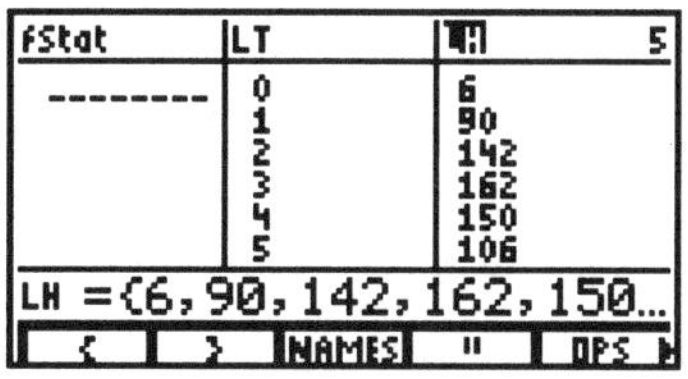

Figure 7.3 Entering data in columns by using 2nd_STAT EDIT.

when you enter the name in the heading the data list will appear in the column.

► *Tip: If a list has values in it, you can delete the data by using the up arrow to highlight the list name and pressing* **CLEAR ENTER**. *Contrary to what you may expect, if you highlight the list name and press* **DEL**, *then the list will disappear from the* **STAT EDIT** *columns, but the data list will still be in memory.*

Finding change in a list: Deltalst(

Figure 7.4 Using Deltalst(to find the average velocity from two data lists.

Returning to the task at hand, we find the average velocity for the first second:

$$(90-6)/(1-0) = 84$$

The TI-86 will do this repetitive calculation and display the whole list at once. We need the function Deltalst(. This is in the catalog or may be pasted from the 2nd_LIST OPS menu after using MORE MORE to show the item Deltal above F4.

► *TI-85 Tip: There is no* **Deltalst(** *command for the TI-85, but you can create it by using a somewhat longer command for* **Deltalst(LH)** *:*
    seq(LH(I+1)-LH(I),I,1,(dimL LH)-1,1)

The Deltalst(command, shown in Figure 7.4, finds the successive differences of the two lists and then divides the lists term by term. The full answer list can be seen by horizontal scrolling.

What does the limit mean graphically and numerically?

The limit is at the foundation of calculus. For example, the key to calculating an instantaneous velocity is to let the time period become closer and closer to zero. Notice that the time period is in the denominator of the average velocity, so letting it reach zero would mean dividing by zero — a definite error. But limits avoid that problem: the idea is to see if the function approaches some value L as the x-values get closer to zero (without ever reaching it). If so, in mathematical notation we write

$$\lim_{x \to 0} f(x) = L$$

The graphical approach to limits

Let's see how the limit works with two particular functions. The first function, f, has no denominator and is defined for all x, even at $x = 0$. The second function, g, is not defined at $x = 0$. (This function is defined in a funny way with $(1)^3$ to put it in the form of a difference quotient, commonly used in calculus.)

$$f(x) = x^2 + 1 \quad \text{and} \quad g(x) = \frac{(1+x)^3 - (1)^3}{x}$$

Now look at the y-values of $g(x)$ for x-values on each side of zero. As shown in Figure 7.5, one is 2.9526... and the other is 3.0478..., so it doesn't take a rocket scientist to figure out that the values are getting close to 3.

This is what we mean when we write $\lim_{x \to 0} g(x) = 3$.

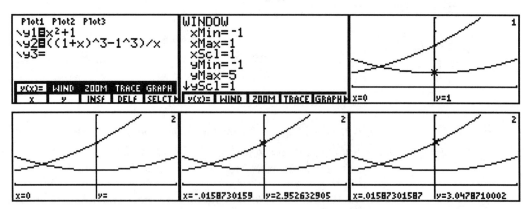

Figure 7.5 When the trace cursor is shown at $x = 0$, the first function is 1. The trace cursor for the second function, however, is undefined at $x = 0$: the y-value is blank. Notice that y-values on either side of the y-axis are very close to 3. This is the graphical meaning of the limit.

The numerical approach to limits

Now let's use the table approach in Figure 7.6 to see the same thing. We press **TABLE** and **TBLST** to do a kind of zoom in close to zero. We get closer and closer by successively using ▵**Tbl** = 1, .1, .01, etc. At zero the second function is undefined, as can be seen by the **ERROR** entry.

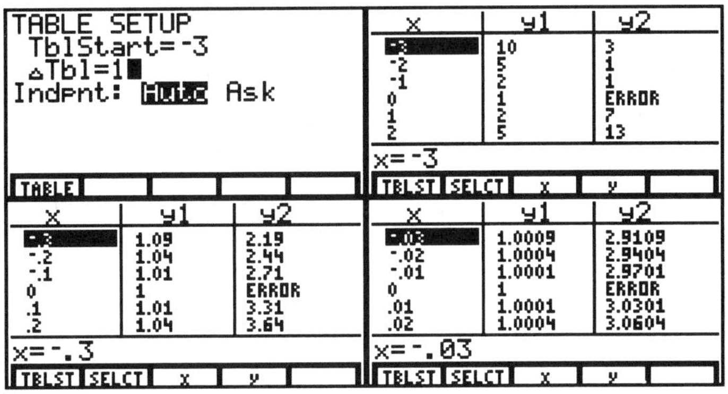

Figure 7.6 Table values close to zero, indicating that the **y1** *values approach 1 and the* **y2** *values approach 3. Use* **TBLST** *to reset with* ▵**Tbl=1, .1, .01** *and the* **TblStart** *values shown.*

The function sin(x)/x

A classic difference equation can be simplified as

$$f(x) = \frac{\sin x - \sin 0}{x - 0} = \frac{\sin x}{x}$$

and we suspect from numerical evidence that

$$\lim_{x \to 0} f(x) = 1$$

To see this, we first defined this new function as **y1= sin(x)/x**. Now, as shown in Figure 7.7, we set **Indpnt:** to **Ask** in the **TBLST** menu, so we can individualize the set of x-values. We then look at values closer and closer to zero.

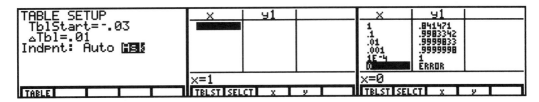

Figure 7.7 Sneaking up on an undefined point (x=0) of sin(x)/x by using various values.

An unreliable table

Let's look at an example where our suspicion is wrong. In Figure 7.8, we define a new function, $g(x) =$ `y2 = sin(2π/x)`, and then using the same five x-values as in Figure 7.6, we might suspect that

$$\lim_{x \to 0} g(x) = 0.$$

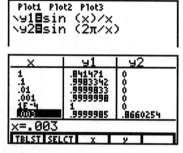

Figure 7.8 *The first five table values for* `y2` *are unreliable for guessing the limit.*

This function is similar to the one used for the 'inaccurate graph' from Chapter 4. Its graph hints that it has no limit. Enter a value that is not the reciprocal of an integer, such as .003, and you see that the function is not going to zero. The limit of $g(x)$ as x approaches 0 does not exist.

Speeding ticket: the Math Police let you off with a warning

A warning: The graphic and table evidence may be very strong to indicate what the limit value should be, but don't make a speedy decision, as this can lead to the wrong conclusion. Only by using careful mathematical analysis can you really prove that the limit exists. For example, we rewrite

$$g(x) = \frac{(1+x)^3 - (1)^3}{x} = \frac{(1 + 3x + 3x^2 + x^3) - 1}{x} = 3 + 3x + x^2$$

and see that

$$\lim_{x \to 0} g(x) = \lim_{x \to 0} (3 + 3x + x^2) = 3$$

Other techniques are necessary for trigonometric functions.

8. FINDING THE DERIVATIVE AT A POINT

The derivative at a point is the instantaneous rate of change that we mentioned in the previous chapter. The average rate of change in miles per hour for a trip in your car is calculated by dividing the distance by the time. The instantaneous rate of change is what is shown continuously by the speedometer as you drive. When you glance down to look at your speedometer, you are finding the of distance with respect to time at that instant. In this chapter we will look at a graphical interpretation of the derivative at a point and learn two ways to use the TI to calculate this value.

Slope line as the derivative at a point: TANLN

In Chapter 4, we saw that we could set a center point and zoom in for a microscopic view of a graph. By continued zooming, the graph began to appear linear. If we can zoom in until the graph appears linear, then we are essentially seeing the tangent line to the graph. The slope of this tangent line to the graph is the derivative at the center point. You can use the MATH submenu to draw a tangent line to a graph at some point, as detailed below.

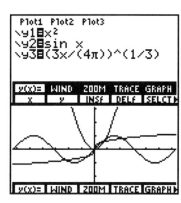

Figure 8.1 Three classic functions for investigating the derivative.

We will consider three classic functions in this chapter:

$$f(x) = x^2, \ g(x) = \sin(x) \ \text{and} \ r = h(V) = \left(\frac{3V}{4\pi}\right)^{\frac{1}{3}}.$$

The last function is the radius of a sphere in terms of its volume. The three functions have been graphed in Figure 8.1 using a ZDECM window setting. Can you see which function is which? (If your sine function is flat against the *x*-axis, then you need to set MODE to Radian.)

Comparing derivatives at a point

At the point $x = 1$, which function has the greatest rate of change?

Let's look at the parabola, y1, first. Turn the other functions off (recall we use SELCT from the y(x)= submenu.). To draw a tangent line and display its slope requires using our slogan GRAPH MORE MATH to put us in the MATH submenu. From there it is MORE MORE to TANLN (F1). Now select the point where you want a tangent line by arrowing to it or by entering its *x*-value. Press ENTER. See Figure 8.2.

Notice that the tangent line is drawn and that the slope value (*dy/dx*) is given at the bottom of the screen. The slope of the tangent line (or the derivative of the function) at the point with $x = 1$ is 2.

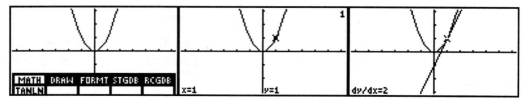

Figure 8.2 *Using the* MATH *submenu item* TANLN *to find a tangent line.*

▶ Tip: *If you want to draw a second tangent line, the first one will remain on the screen unless you use* GRAPH DRAW CLDRW *to start with a fresh graph. Since you have to use* MORE *to find* CLDRW, *it is often easier to go to the function definition screen (*y(x)=*) and retype any character. Then* GRAPH *will start with a clear screen (without the first tangent line).*

The derivative at a point, without the tangent line: dy/dx

Tangent lines are like training wheels on a bicycle: they eventually become unnecessary. We now show how to just find the derivative at a point. In Leibniz notation, *dy/dx* is the symbol for the derivative. This selection is available in the MATH submenu. Figure 8.3 shows finding the derivative of the sine function at $x = 1$. This derivative value, about one-half, is less than the derivative of $f(x) = x^2$ at 1; in comparison, the sine values are increasing more slowly near this point. Our next example shows a non-graphical approach to finding the derivative.

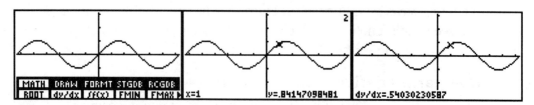

Figure 8.3 *Using* MATH dy/dx *to find the derivative at a point on the graph.*

Using a numerical approach to finding the derivative at a point: `der1(...)`

From the graph in Figure 8.1, we saw that the function y3 was increasing the least at the point $x = 1$. From the home screen we can find the derivative of this function (y3) by using a 2nd_CALC menu command,

$$\text{der1(y3,x,1)}$$

The basic calculus commands are in this menu (CALC), but they don't look like any standard calculus notation. Let's translate: der1 as you may have guessed, is an abbre-

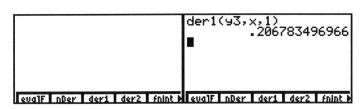

Figure 8.4 Using `2nd_CALC der1()` *to find the derivative of a function at a point.*

viation of 'derivative 1,' i.e., the first derivative and der2 is the second derivative. The command nDer we will be explained shortly and fnInt will be used heavily in Part III Integral Calculus.

The first entry in der1 is the function, the second entry is the independent variable of the function, and the third entry is the point at which to evaluate. The function need not be stored in the y(x)= editor and it need not have x as its variable. For example, you could enter der1(T²,T,1) and get a value of 2 (as we found for y1).

In summary, we have found y1'(1) = 2, y2'(1) ≈ .5, and y3'(1) ≈ .2, which confirms our visually based answer that the derivative of the radius function is changing the least of the three functions at $x = 1$.

Comparing the exact and the numeric derivative

The 2nd_CALC menu offers two kinds of first derivatives: nDer and der1. What is the difference? As we have seen, der1 finds the first derivative of the function. But it has some limitations because it finds values using a formula-based approach (like knowing that $y' = \cos x$ for $y = \sin x$). An abbreviation of 'numerical derivative,' nDer will find values for any function, but in some circumstances it gives invalid results.

How does the TI calculator find the numeric derivative?

You know from your text that the derivative of f at a is defined as

$$\lim_{h \to 0} \frac{f(a+h) - f(a)}{h}, \text{ if the limit exists.}$$

The calculator finds a single value to report as the derivative at point a using the related formula

$$\texttt{nDer(f(x),x,a)} = \frac{f(a+h)-f(a-h)}{2h}, \; h = 0.001$$

The value of h is called δ and can be changed in the **2nd_MEM TOL** screen. (**2nd_TOLER** on the TI-85).

Warning: Approximations by nDer can give false results

It is important to realize that **nDer** is an *approximation*. This short cut can get us into deep trouble with certain points of some functions. For example, we know the function $f(x) = 1/x$ is not defined at zero and thus has no derivative there. If we try using **der1** from the home screen, we get an error message (as we should.) But if we try **nDer** we erroneously get a value of 1 million for the derivative. (From a graph of $y = 1/x$, trying to use **dy/dx** will not allow us to find a numerical derivative at zero.)

```
der1(1/x,x,0)
ERROR 02 DIV BY ZERO
```

```
nDer(1/x,x,0)
                1000000
```

Figure 8.5 At x=0, where the derivative is undefined, **der1** *signals an error, and* **nDer** *reports an invalid value.*

This problem with **nDer** is not restricted to the obvious cases where the function itself is undefined. Consider our function **y3=(3x/(4π))^(1/3)**, it is defined for all real values, and you might expect to find a derivative at $x = 0$ in the same way we did for $x = 1$. However at $x = 0$, the tangent line actually is a vertical line. This means that the derivative is undefined (has no slope) at $x = 0$. Using **nDer** gives the wrong answer, as shown in Figure 8.6. You will find that the **der1** correctly gives an error messages. Does this mean we should never use **nDer**? As we will discuss below, **der1** has limitations, so there are occasions where **nDer** is required — we just need to exercise caution.

```
nDer(y3,x,0)
          62.0350490899
der1(y3,x,0)
```

```
ERROR 02 DIV BY ZERO
```

Figure 8.6 The nDer reports an invalid value when the derivative is undefined but **der1** *gives the correct message.*

Improving the accuracy of an approximation (optional)

The default value δ = .001 can be changed to increase the accuracy of our **nDer** approximation. However, as we try to increase the accuracy by making δ extremely small, we run afoul of the calculator's computational restrictions and our accuracy declines. You can use **2nd_MEM TOL** to reset δ, before calculating **nDer(2^x,x,0)** with each of the successive values:

δ = **.001**, **E-5**, **E-8**, **E-12** and **E-15**

In Figure 8.7 for reasons of screen compactness, we set δ and evaluate **nDer** together on the home screen. Of the δ values, δ = .00001 is closest to the true value of the derivative of $y = 2^x$ at $x = 0$. By using calculus we know the true value is ln(2), which to eleven decimal places, is 0.69314718056.

Figure 8.7 Resetting a default to increase accuracy and going too far. Warning: this is a very dangerous setting for δ. Before proceeding reset δ back to the default δ = .001.

➤ *Tip: Use **2nd_EE** to enter small values. Use negation, not subtraction.*

➤ *Tip: Be careful when finding the derivative at a point: be sure the function is defined at the point, and that the tangent line exists and is not vertical.*

Problems with **der1(...)**

We prefer **der1** to **nDer**, but there are some limitations. First, recall that the calculator must know the derivative of the function for **der1** to work. In the absence of a list of known functions, you should be prepared for an error message if your function is out of the ordinary. For example, the function **int** is not on the list, so the expression **der1(int x,x,.5)** returns an error. But **nDer(int x,x,.5)** gives the correct answer. We are also were unable to use **der1** on a composite function, like **y3=y1(y2(x))**, even when **y1** and **y2** are simple functions.

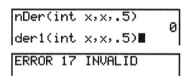

*Figure 8.8 **der1** and its error message when it does not know the differentiation rule.*

A second failing of **der1** is that it can make domain errors. Consider the graphs in Figure 8.9 on a **ZDECM** screen. The **y2=der1(y1,x,x)** should not be defined for negative values (such as $x = -1$) which are not in the domain of **y1**=ln x.

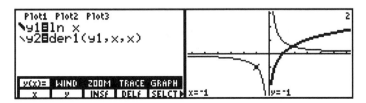

*Figure 8.9 **der1** gives values where the function is undefined.*

9. THE DERIVATIVE AS A FUNCTION

If we know the value of the derivative at a whole set of points, then we can define a new function called the derivative of f:

$$f'(x) = \lim_{h \to 0} \frac{f(x+h) - f(x)}{h}, \text{ if the limit exists.}$$

The derivative at a point was defined in the last chapter and we showed that the `2nd_CALC` menu commands

`nDer(T²,T,1)` and `der1(T²,T,1)`

gave a numerical value for the derivative of $f(T) = T^2$ at the point $T = 1$. We can now find this new derivative of f at any point x by defining

$$f'(x) = \text{der1(T²,T,x)}$$

(We could also use `nDer`, but `der1` tends to have fewer problems.)

In this chapter we will show how to graph this function and better understand the relationship between a function and its derivative.

Viewing a graph of derivative function

We know T is a dummy variable in the above function definition, so why not call it x? We can, at the risk of some confusion. The first time you see `der1(x²,x,x)`, you might think there is an extra x. We started in the last chapter with the function T^2 and variable T to help make it clear here that not all these x's are being used in the same way. In the first two entries of `der1(x²,x,x)`, the x is a dummy variable, while in the third entry, it is the independent variable. This is a confusing aspect of the TI notation.

Matching a function to the graph of its derivative

We consider again the three classic functions from the last chapter:

$$f(x) = x^2, \; g(x) = \sin(x) \text{ and } r = h(V) = \left(\frac{3V}{4\pi}\right)^{\frac{1}{3}}.$$

The three functions were stored in `y1`, `y2`, and `y3` and graphed using a `ZDECM` window setting. Now we graph the derivative function for each of the three functions and display them in Figure 9.1. Can you match the curves to the derivative functions?

9. THE DERIVATIVE AS A FUNCTION

Figure 9.1 Three derivative functions for three classic functions.

We could, of course, graph them one by one, but using a little thought we can identify them just by looking at the features of the graph. Consider the parabolic function *f*: it is decreasing until it reaches zero, then it is increasing. Since the derivative gives the instantaneous rate of change, this means that the derivative values are all negative to the left of the origin and are all positive to the right of the origin. Of the three options, this describes the line. (You may also know the power rule of derivatives which says that the derivative of a quadratic function is a linear function.) Now consider the sine function: it oscillates between decreasing and increasing, so the derivative should oscillate between negative and positive. There is only one function that does this and it looks like a cosine function (another rule you may know). The remaining derivative function is always positive and has a spike at zero; this fits the slope patterns of the y3 graph. As a reminder, the derivative function of *h(x)* is undefined at zero, but nDer erroneously gives a value there.

Graphing a function and its derivative in the same window

It is instructive to graph a function and its derivative in the same window, but you may want a means of distinguishing which is which. This is accomplished by using the

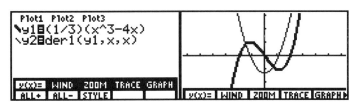

Figure 9.2 Two styles to distinguish a function and its derivative function in the same window.

STYLE feature in the y(x)= submenu. The slash sign that you have seen to the left of the yi= is the icon for a regular line. (This feature is not available on the TI-85.) By pressing STYLE (which is F3 in the second set of five options under y(x)=), you can change the graphing mode a bold line. Unfortunately, to change back to the regular style, you must press STYLE six times to cycle through the entire set of styles. See Figure 9.2

Functions and their derivatives do not always fit very well in the same window. Consider a slight modification to the above example and add 100 to the y1 function, shown in Figure 9.3. The function graph will not appear in the ZDECM window, but the derivative function will be identical to the derivative function shown in Figure 9.2.

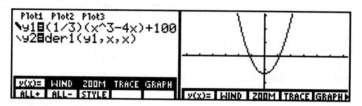

Figure 9.3 *The function graph and derivative graph rarely fit in the same window.*

There is a lesson here. Showing the graph and its function in the same window is a parlor trick and must be carefully designed to work. Further, it should be realized that a function serving as a model and its derivative will use different interpretations of the *y*-axis. For example, when modeling motion, the distance function might be in the feet and the derivative function would be in feet per second.

The derivative function using the math definition

Let's suppose we had a lame calculator that did not have the derivative feature built in. We can make one by hand, based on the limit definition of the derivative:

$$f'(x) = \lim_{h \to 0} \frac{f(x+h) - f(x)}{h}$$

As an example, define a new function y1 = 1/x² and leave y2=der1(y1,x,x) in place from the previous example so that we won't need to reconstruct it later. We construct a handmade function y3 by using y1 and a very small *h*; let *h* = .001. In Figure 9.4, these definitions are shown on the home screen so that the full definition can be more easily seen. Change the y3 style to have a leading circular cursor (the fourth option). After pressing ZDECM we can watch the circular cursor trace along our derivative function. They show a very close match the derivative is positive and increasing to the left of the origin, negative and increasing to the right.

▶ *Tip: The value h which we used is the default setting for δ used when calculating* nDer. *Both quotient methods have the same limit value when h goes to zero, but the TI's version is more accurate when h is given a specific value. See page 53 for details.*

Figure 9.4 A handmade numerical derivative function graphically compared to the true derivative.

The function that is its own derivative: $y = e^x$

The graphs of the function `y1` = `1/x²` and its derivative in Figure 9.4 are similar for negative values of x. Could some function be its own derivative? To save time in guessing, we will try an exponential function. In Figure 9.5, we first change the function `y1` to be our old doubling function $y = 2^x$ and graph it along with its derivative (leave `y2` and deselect or clear `y3`). However, we will use our new style feature to have a leading circular cursor. This helps visually distinguish between two functions that have close values.

Now change `y1` to the tripling function $y = 3^x$ and graph again. This function is very close to its derivative. The exponential function we are looking for has a

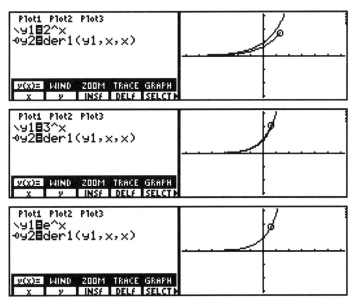

Figure 9.5 Looking for a function that is its own derivative.

base between 2 and 3. In calculus, we find that this amazing number is 2.718..., an irrational number denoted by e.

Using lists to estimate a derivative function: Deltalst(

Sometimes discrete data is gathered and we use it to construct the average rate of change between successive data values. For example, consider the table in Figure 9.6 showing distance traveled by a car.

We can use **2nd_STAT EDIT** to enter the times and distances as list in **LT** and **LD**, respectively. Next apply **2nd_LIST OPS Deltalst(** to the list and then divide by **Deltalst(LT)**. (Since the common difference between measurement times is equal, you could just divide by 3 in this case.)

Store these difference quotient values in **LDQ**, as shown in Figure 9.7. However, the 20 at the top of the list **LDQ** is our estimate for time 3 seconds, which is the second time entry. Thus we need to shift the list down by one place.

Time (seconds)	0	3	6	9	12
Distance (meters)	0	60	159	288	441

Figure 9.6 Distance traveled by a car.

Although technically we don't know that the car had velocity zero at time zero, we make that assumption. Adding 0 to the list is most easily done in the list editor. Press **STAT EDIT** and see the list edit screen. Move to the empty heading and enter **LDQ** to paste the list in the third column. Arrow down to the first entry and press **2nd_INS** to place a zero at the top of the list. In this example, we interpret the values shown in **LDQ** as the velocities.

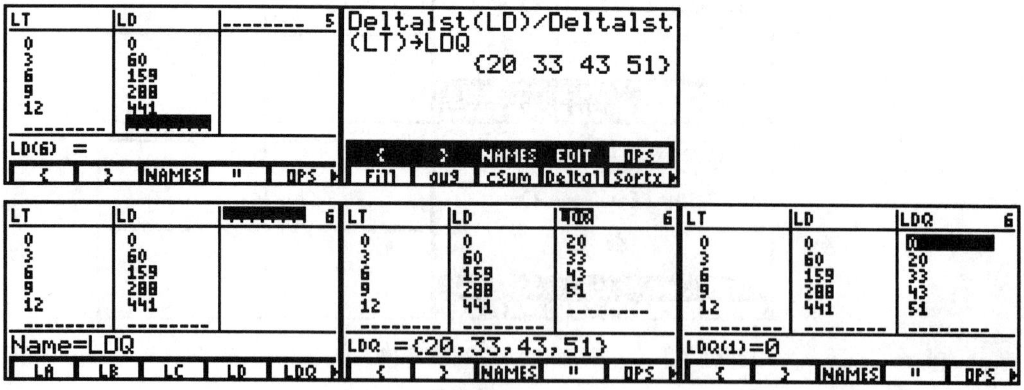

Figure 9.7 Making a list of numerical derivatives.

10. THE SECOND DERIVATIVE: THE DERIVATIVE OF THE DERIVATIVE

The derivative function for the function y1 was created by setting
$$y2 = der1(y1,x,x)$$
Since y2 is a function itself, there is no stopping us from writing the *derivative of the derivative* as a function:
$$y3 = der1(y2,x,x)$$
But we have a built in function for the second derivative and we write
$$f'(x) = der1(y1,x,x) \text{ and } f''(x) = der2(y1,x,x)$$

In this chapter we will see how the second derivative tells us about the concavity of the graph of the original function.

How to define and graph f(x), f'(x), and f''(x)

Let's start with a known result. If $f(x) = x^2$, then by the power rule $f'(x) = 2x$ and $f''(x) = 2$. We can see this clearly in Figure 10.1.

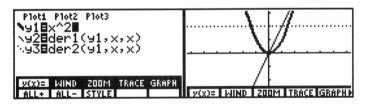

Figure 10.1 A graph of $y = x^2$ with its first and second derivatives.

Graphs using der1 and der2 (and even more so nDer) in their definition require calculations that slow the process consider-ably. By setting xRes = 3, only one-third of the usual *x*-values will be evaluated, thereby increasing the speed. Another benefit from this is that the dot style of the second derivative graph will be more pronounced. (Recall that xRes and STYLE are not available on the TI-85.)

➤ *Tip: If the graph is taking a very long time, press* ON *to stop it and reset* xRes *to improve the speed.*

We have commented that, in general, trying to graph a function and its derivative in the same window is not practical. We see this again in the next example as we analyze the logistic function and its derivatives.

Looking at the concavity of the logistic curve

In most growth situations, a logistic growth model makes more sense that a pure exponential model. A new software company may keep doubling employees, but the growth has to slow down or else, like the folding of the paper that reached the moon, the number of employees would eventually be out of this world, literally. Suppose the logistic function entered as y1 in Figure 10.2 gives the number of employees in the company at time x.

Figure 10.2 Defining a logistic function and its first and second derivatives.

We want to view the first and second derivatives and see what they tell us about the function. It is necessary to consider each one individually and find an appropriate window for each graph. The graph y1 has a range of values that would make seeing y2 and y3 impossible in this window.

The function y1 is monotonically increasing on the interval shown in Figure 10.3, so the derivative function y2 must be positive. Notice that y2 peaks and decreases to zero. The peak is at the point of the fastest growth of y1. The second derivative shows the rate of change of the rate of change. We see that the value of y3 is zero at the peak of y2.

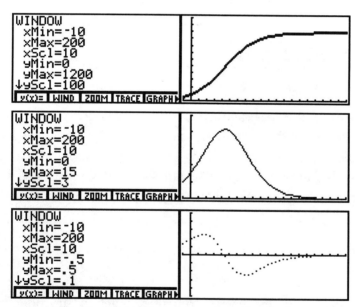

Figure 10.3 The logistic function and its two derivative functions.

In this example we can see that the second derivative (y3) is zero at about 40. This is because the growth rate (y2) peaks at about 40 and begins to slow. On the graph of y1, this is the point of fastest growth. It is a point where the concavity is changing from positive to negative. We call this a point of inflection on the graph. Since this is an important point on y1, we want to identify it more exactly.

10. THE SECOND DERIVATIVE: THE DERIVATIVE OF THE DERIVATIVE

Our way of identifying this point is to find the zero of the second derivative. Recall this is **GRAPH MORE MATH ROOT**. The computation is shown in Figure 10.4, but be warned that it will take several seconds for your calculator to do this. Finding the zero of the second derivative would be much faster if an actual algebraic formula were known.

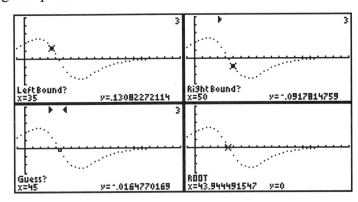

*Figure 10.4 Using **ROOT** to find the zero of the second derivative.*

Creating a numeric second derivative using a table

We can find the derivative of the velocity function in the same way we found the derivative of the distance function. The second derivative of the distance function is the acceleration. Recall from the last chapter how we the calculated a car's velocity at five different times. We return to these lists and use the list editor for convenience. Press **2nd_STAT EDIT** to begin where we left off. Move up to the heading row and over an empty heading where we name our list: press

ENTER LDQ ENTER

Next we define how the list is built by entering our formula. Paste the **Deltalst(** command either from the catalog or from the **LIST OPS** submenu (**Delta1** in the third group of five). The average derivative of the average derivative is calculated in **LSDQ** as shown in Figure 10.5.

What does acceleration mean? Although the velocity is increasing, it increases by less and less each three second interval. So the acceleration, the rate of change of velocity, is decreasing. Let's face reality: a car in low gear can peel out of a parking lot and make a squeal, but at higher speeds there will be less acceleration.

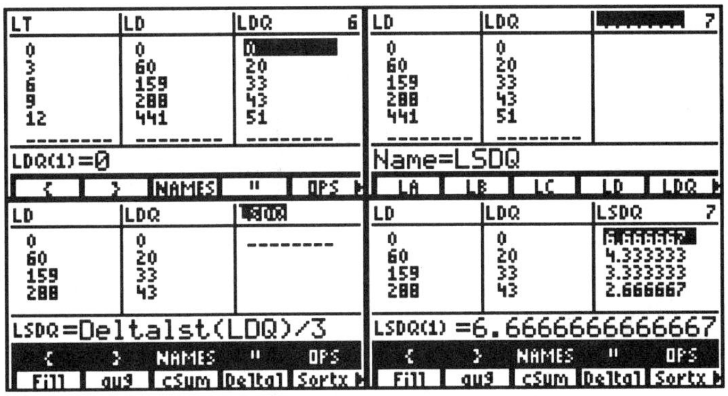

Figure 10.5 A discrete list of distance, velocity and acceleration values over time.

11. THE RULES OF DIFFERENTIATION

Using the definition of the derivative to find a derivative function is cumbersome; fortunately there are shortcuts to finding derivative functions. We will use the calculator to see examples of the main three rules. These rules must be proved analytically, but a graphical verification gives us added confidence in them. Also, we can show graphically that some common guesses for the rules are wrong.

The Product Rule: $(fg)' = f'g + fg'$

We take two common functions and graph their product and the derivative of the product, as shown in Figure 11.1. We see from the graph that the product function has a local maximum at $x = -2$, and thus the derivative function should be zero there.

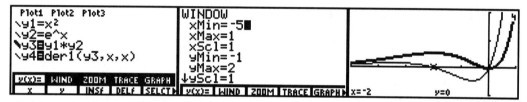

Figure 11.1 A product function and its derivative. Use **TRACE** to locate a zero of **y4**.

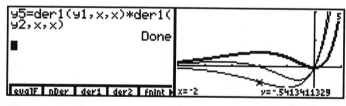

Figure 11.2 The product of the derivative functions does not have a zero at the product function maximum, therefore $(fg)' \neq f'g'$.

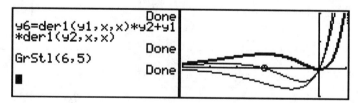

Figure 11.3 The correct derivative formula function, **y6**, traces over **y4**, the derivative of the product.

Now let's suppose our guess for the product rule is that that derivative of the product is the product of the derivatives. This is not such a silly guess since the derivative of a sum is the sum of the derivatives; but it is wrong. We can see this in Figure 11.2 by graphing the product of the derivatives. The product of the derivative functions does not have a zero where it should, at $x = -2$, so it is not the derivative of the product function.

To see that the product rule holds for these two functions in this window, we enter the true formula with a leading cursor style in Figure 11.3 in order to see that the y6 graph traces over the y4 graph. (TI-85 users will need to use trace on the two curves to see that they are the same.)

The Quotient Rule: (f/g)' = (f'g-fg')/g²

Let's check the quotient rule in the same way. This means we can recycle the functions, make a minor change to the window setting, and get the graph shown in Figure 11.4. Notice the two zeros of the derivative function are at the local maximum and minimum of the quotient function.

We can again make a feasible but incorrect guess, namely that the derivative of the quotient is the quotient of the derivatives. We see our folly easily since the quotient of the derivatives is not zero as $x = 2$. Now enter the correct formula in y6 and see that it traces over y4.

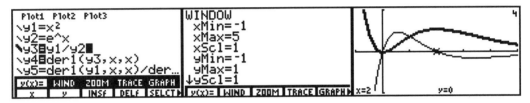

Figure 11.4 A quotient function (bold). Its derivative, y4, has a zero at the max of y3.

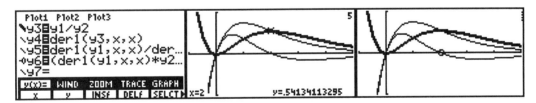

Figure 11.5 The incorrect quotient function, y5=der1(y1,x,x)/der1(y2,x,x), has no zero at x=2; it does not trace y4. But the correct formula, y6=(der1(y1,x,x)*y2-y1*der1(y2,x,x))/(y2)², does trace the derivative of the quotient function.

The Chain Rule

Thinking of a function as a composite function and using the chain rule will often simplify finding the derivative of a function. Consider $y = (x^2+1)^{100}$. A straightforward but impractical approach would be to expand the expression, write it as a polynomial of degree 200, and then differentiate term by term. Instead, we apply the chain rule and find the derivative quickly and easily:

$$y' = 100(x^2+1)^{99}(2x) = 200x(x^2+1)^{99}$$

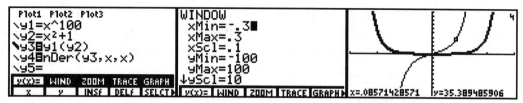

Figure 11.6 A composite function and its derivative.

Figure 11.6 shows the graphs of y and y' (notice the extreme scale difference between x and y). This is an instance where `der1()` can not be used; it would correctly graph the derivative of the function `y3` when defined as `y3=(x²+1)^100`, but not when defined in the symbolic form `y3=y1(y2)`.

Some care should be exercised in entering the function for the chain rule formula. The derivative of the outside function is evaluated in terms of the inside function, so the outside function is evaluated at `y2`. Hence for this calculator we write

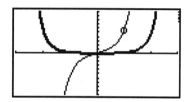

Figure 11.7 The correct formula, `y5`, traces the derivative of the composite function.

`y5=nDer(y1,x,y2)*nDer(y2,x,x)`

and graph with the leading cursor style. We see in Figure 11.7 that the graphs of the two derivative functions are the same.

The derivative of the tangent function

Recall that the window used for graphing trigonometric functions is often crucial. In Figure 11.8 we first use the `ZOOM ZSTD` window, then the `ZOOM ZTRIG` window, to view `y1 = tan(x)`. This should remind you that x-values are sampled evenly across the window and the function values at these points are then connected to form the graph. It is clear from the graph of the tangent function that it is an increasing function within an interval such as $-\pi/2 < x < \pi/2$ and that it is undefined at multiples of $\pi/2$. Thus we expect the derivative to be positive between $-\pi/2 < x < \pi/2$ and undefined at multiples of $\pi/2$.

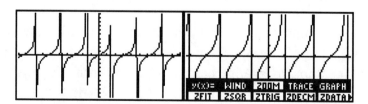

Figure 11.8 Graph of `y1 = tan(x)`, first with `ZSTD` and then with `ZTRIG`.

We can find the derivative of the tangent function from the quotient definition $\tan(x) = \sin(x)/\cos(x)$. Set $y = \tan(x)$ and use the quotient rule to derive $y' = 1/\cos^2(x)$. This is always positive and is undefined at multiples of $\pi/2$ (where the cosine is zero).

In Figure 11.9, we check that `y2=der1(y1,x,x)` coincides with the graph of the algebraic derivative, `y3 = 1/(cos(x))`². We see that `y2` correctly evaluates the undefined points and does not connect the function values over points of discontinuity.

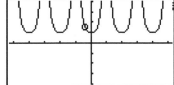

➤ *Tip: The conventional mathematical way of writing a power of a trigonometric function, such as $\cos^2(x)$, gives a syntax error. Use $\cos(x)^2$ or, better yet, $(\cos(x))^2$.*

Figure 11.9 Graph of the algebraic derivative of tan(x) with the **der1()** derivative tracing over it.

Notes on nDer and xRes

If you enter the numeric derivative of `y1` in `y4` and graph it, you may be surprised: see Figure 11.10. This creates double vertical lines between the defined intervals. What is going on? If you trace to a value like $\pi/2$, as shown in the second panel, then you will see that the numeric

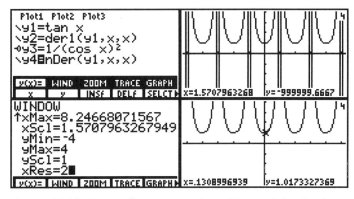

Figure 11.10 Reset **xRes** to avoid graphing undefined values.

derivative has been calculated incorrectly as a large negative number.

The moral of this graph is that we must be constantly vigilant in believing the numeric derivative function at points where the function is undefined. A cosmetic remedy to this problem is to make the sampling avoid the bad places. In this example we can raise the **xRes** setting as shown in the bottom of Figure 11.10 and it won't evaluate at multiples of $\pi/2$. (Remember, there is no **xRes** setting on a TI-85.) An **xRes** setting of 2, as in Figure 11.10, will skip every other *x*-value that the calculator normally would use in its sample of *x*-values between **xMax** and **xMin**. These 127 samples are formed by adding multiples of (**xMax** − **xMin**)/126 to **xMin**. The only **ZOOM** setting that changes **xRes** is **ZSTD**; it resets to **xRes=1**.

12. OPTIMIZATION

One of the powerful uses of the derivative function is to help find the maximum or minimum of a function. But we must confess that calculators and computers with graphing capabilities can, in most cases, find maximum and minimum values of a function without your having to know anything about calculus. In this section there are examples showing both the calculus and the non-calculus approaches.

The ladder problem

Typically, optimization problems arise from real-world applications. The 'ladder' problem is to determine the longest ladder that can be carried horizontally around a corner that joins two hallways. We will assume that the hallways are different widths: the narrower one is 4 feet and the wider one is 8 feet wide. Figure 12.1 shows the position where the ladder could get stuck: it touches both walls and the corner. For the ladder to make this corner, it needs to fit in the hallway for every angle θ, even this tightest one. This leads to minimizing the equation

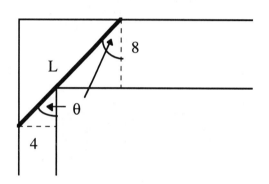

Figure 12.1 *A ladder carried horizontally around a corner.*

$$L = f(\theta) = 4/\sin\theta + 8/\cos\theta.$$

A ladder of that minimal length will fit for all angles, and it will be the longest possible ladder that works since it touches both walls at the tightest angle.

Finding such an equation is the hard part. This equation was derived by thinking of the ladder as being divided at the corner point and using the triangle trigonometry definitions for the sine and cosine.

To find the minimum value of L, follow these steps:
- be sure that the **MODE** setting is **Radian**
- enter the function L in **y1** with x as the variable
- set the window.

You might be tempted to use **ZTRIG** but, like most models, there is a more restricted domain in this example. The angle θ must be greater than 0 but less than $\pi/2$. To set the window y-values, we can be generous and say that the minimum ladder will be under 50 feet. Use **TRACE** to approximate the lowest point on the graph, as shown in Figure 12.2.

12. OPTIMIZATION   69

*Figure 12.2 A minimum found by using **TRACE** and the trial-and-error method.*

Using the **FMIN** feature to find the minimum

We can use **GRAPH MORE MATH FMIN** to directly find $y \approx 16.65$. You must provide a Left Bound, a Right Bound, and a Guess. The *x*-values must be within the viewing window. These values can be entered directly on a TI-86 (but not the TI-85) or by using the arrows and pressing **ENTER**.

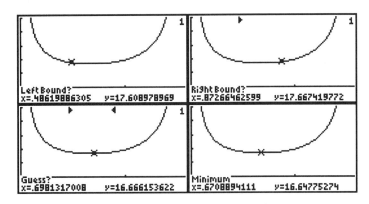

*Figure 12.3 A minimum found by using **FMIN**.*

Using the graph of the derivative to find the minimum

Calculus tells us that the local maximums and minimums occur at points where the first derivative is zero or undefined. Using styles (TI-86 only) to distinguish between the curves, we see the function (bold) and its derivative in Figure 12.4. The second derivative is not visible except for a few dots at the top of the screen.

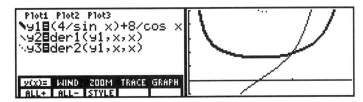

Figure 12.4 A minimum found by graphing the function and its first and second derivatives.

To find the minimum, trace to the closest point to the zero of the first derivative and use the up arrow to read the function value. Use the up arrow again and the value of the second derivative is shown as $y \approx 50$; since this is a positive value, the point is a minimum by the second derivative test.

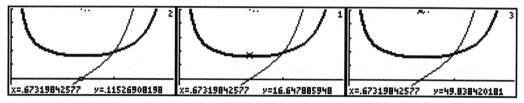

Figure 12.5 A minimum found by using an approximate zero of the derivative function and checking the second derivative to confirm whether it is a maximum or a minimum.

After all of that, you may wonder why we bother to use the derivatives. One reason is that, in many cases, we can use derivatives to find an exact answer. Such 'closed form' solutions can be important for using in other calculations or for insuring unlimited accuracy.

Box with lid

Suppose we have an 8.5 x 11 inch sheet of paper and want to cut squares and rectangles from the corners to create a folded box with lid. See Figure 12.6. We want to maximize the volume. Notice that if the x cut is very small then the box will be so shallow that it hardly holds anything. If the x cut is large, then the bottom, **b**, is so small that the box again holds very little. We first find a volume function that depends on the length x of the cut-out square's side:

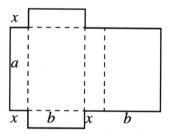

Figure 12.6 A diagram of cutting corners to create a box with lid. Fold on the dotted lines.

$$V = (8.5 - 2x)(11/2 - x)x$$

Now we will show three different ways to solve this problem. A precalculus approach is shown first.

Using the TRACE to find the maximum

Figure 12.7 shows the screens to obtain an approximate maximum by using TRACE. The derivative definitions y2 and y3 which reference y1 are kept for later analysis but they are deselected for the first graph.

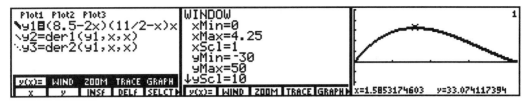

Figure 12.7 A maximum found by using **TRACE** *and the trial-and-error method.*

Using the zero of the derivative to find the maximum

Next we use calculus. Select all three functions and change **xRes** to speed up the graphing process. This also makes a dotted style graph more spread out. The zero of the derivative function gives us the size of cut needed. In this case, we cannot trace to an exact zero, but use the *x*-value for *y* closest to zero. ($x \approx 1.62$) We can see that the second derivative is negative at this point, which confirms that this is a maximum rather than a minimum.

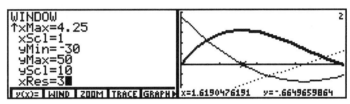

Figure 12.8 A maximum found by using the approximate zero of the derivative and seeing that the second derivative is negative.

Using SOLVER and the der1 to find the maximum

We can improve on the accuracy of the above answer by having **SOLVER** find the zero of the derivative. Recall that the procedure is to enter the equation, then put the cursor on the variable you wish to know and press **SOLVE**. The current value (**x=1.619...**) is treated as a guess. The calculated answer will be shown with a box beside it. This answer is stored in the variable so we can use it in other calculations. In Figure 12.9 we find the maximum volume using the volume function **y1** and then check the second derivative **y3** to be sure it is negative, signifying a maximum. When using a graphical method it is obvious from the graph whether it is a minimum or maximun, but with a numerical method it is not as obvious and checking the second derivative takes on added importance.

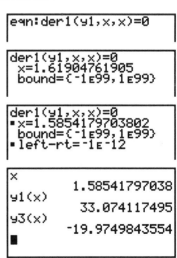

Figure 12.9 A maximum found by using **SOLVER** *to give the zero of the derivative and substituting back.*

Using the second derivative to find concavity

An important function in statistics is the *standard normal distribution* function. It has a graph that is bell-shaped and has the definition

$$f(x) = \frac{1}{\sqrt{2\pi}} e^{-x^2/2}$$

For simplicity we define a function in the same family that does not have the coefficient $1/\sqrt{2\pi}$. The equation and (bold) graph are shown in Figure 12.10. The graphs of the first and second derivatives have been included.

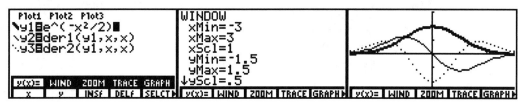

Figure 12.10 A bell-shaped curve and its derivatives.

The concavity of the normal curve is revealed by finding the zero of the second derivative. The zero with positive x in this example gives a point where the second derivative changes from negative to positive. That is where the function changes from being concave down to concave up. The zero with negative x gives a point where the function changes from being concave up to concave down. An important characteristic of this bell-shaped curve and its multiples such as the standard normal distribution is that they have points of inflection at $x = \pm 1$.

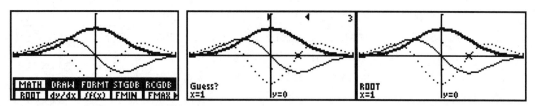

Figure 12.11 A point of inflection for this bell-shaped curve is at $x = 1$. (The bound setting screens are not shown.)

Finding a point of inflection: INFLC

Among the features of the GRAPH MORE MATH submenu is (in the second five) INFLC (F1) which will find a point of inflection without requiring you to define and use the second derivative. The Left Bound, Right Bound, Guess sequence is used to isolate which point of inflection you want to find. In Figure 12.12, the facing screen arrows in the middle panel show the Left and Right Bounds and a Guess that looks close to the positive inflection point. The final screen shows that we have an inflections point at exactly $x = 1$.

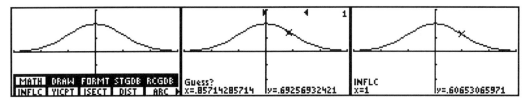

Figure 12.12 Finding a point of inflection using INFLC.

PART III
INTEGRAL CALCULUS

13. LEFT- AND RIGHT-HAND SUMS

14. THE DEFINITE INTEGRAL

15. THE FUNDAMENTAL THEOREM OF CALCULUS

16. RIEMANN SUMS

17. IMPROPER INTEGRALS

18. APPLICATIONS OF THE INTEGRAL

13. LEFT- AND RIGHT-HAND SUMS

The fundamental activity of integral calculus is adding. In the discrete case, we sum a set of values. In the continuous case, we use the integral to sum over an interval. In this chapter we will restrict our attention to finite discrete sums. You probably already know that these sums are approximations to the value of a definite integral. We will make that connection in Chapter 16, Riemann sums.

Distance from the sum of the velocity data

Time (sec)	0	2	4	6	8	10
Velocity (ft/sec)	20	30	38	44	48	50

Figure 13.1 Velocity of a car every two seconds.

If you drive 50 miles per hour for 3 hours, then you will have traveled 50 + 50 + 50 = 150 miles. If you drive 20 mph for two hours, then 30 mph for two hours, and finally 40 mph for two hours, then in the six hours you will have traveled 20(2)+ 30(2) + 40(2) = 180 miles. We rarely travel a constant speed. Figure 13.1 shows velocity readings at 6 different times. We do not know if we traveled mostly at 20 ft/sec or 30 ft/sec for the first two seconds. If we assume the velocity is constantly increasing, then these two numbers give us lower and upper bounds for the first two seconds.

We simply add the first five and multiply by 2, and then repeat this with the last five velocities to find a lower and upper bound on the distance traveled in ten seconds. The TI has list variables (named like any other variables) to do this job. To distinguish list variables, in this book we will start their names with the letter L. The set symbols { and } used in defining lists are found in the 2nd_LIST submenu; you may also want to put them in CUSTOM for convenience.

▶ *Tip: If you make long names, then you will probably want to paste them from a menu of list names as you use them in expressions. It is convenient to use all upper case letters in a name since the STO→ key puts you in ALPHA-lock.*

Creating and summing lists from the home screen: {...}→L

After entering the first data list and storing it in **LLB** (see Figure 13.2), we can use **2nd_ENTRY** to edit and create the second list from the first. Delete the first entry, 20, insert 50 at the end of the list, and change **LLB** to **LUB**. The sum command can be typed or pasted from either **2nd_MATH MISC** or **2nd_LIST OPS MORE**.

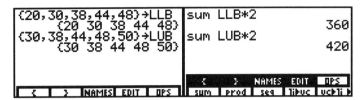

Figure 13.2 Finding lower and upper bounds on distance traveled.

➤ *Tip: By using **2nd_LIST OPS MORE**, you can have both **sum** and the set brackets visible on menus.*

Creating and summing lists from 2nd_STAT EDIT

The alternate way to enter lists, which we used in Chapters 9 and 10, is the column list editor. (Reminder: This feature is not available on the TI-85.) It is particularly handy when the list is long and when you need to edit the data. Press **2nd_ STAT EDIT** to reach the list editor screen. You can go to an empty heading to name it and then enter data in that column.

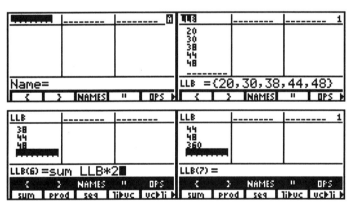

*Figure 13.3 Press **DEL** to remove previous lists. Paste or enter an existing list name into a column heading (or enter a new name and new data.) Arrow to the bottom of the list. Enter **sum** from the **OPS** submenu (which requires **MORE**) and then **LLB*2**. Press **ENTER** to see the calculation.*

You can also paste or type in the name of a pre-existing list and the data will appear in the column below. To remove a list from the screen editor, press **DEL**. In Figure 13.3, since we have already entered the data, we delete any previous lists and paste in the list name **LLB**. The computation is done in a cell below the data, but it could be done in any cell.

Using sequences to create a list of function values: seq(...)

The TI has a very handy feature called seq. Essentially, the sequence command makes a list of values using a given expression with the index taking values from a starting point to a stopping point, increasing by a given increment:

seq(*expression, index, start, stop, increment*)

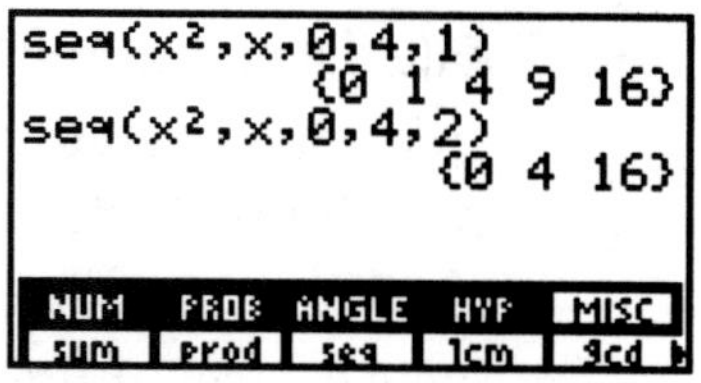

Figure 13.4 Using seq() *from* 2nd_MATH MISC *to create a list of square values, then a set of even squares.*

As an example, we can create a list of squares from 0 to 16. We first use the squaring function inside the sequence command, as shown in Figure 13.4. The TI-86 default setting for the increment is 1 and this can be left off. However, entering the increment is mandatory for the TI-85.

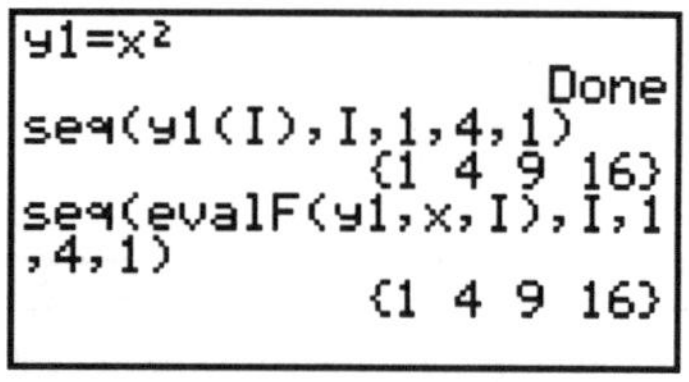

Should you want the list of function values for a named function, say y1=x^2, there are two alternatives as shown in Figure 13.5. The first alternative is the most natural, following the format above, but is only valid for the TI-86. In the second method, which is valid for both models, the command evalF(is used from the catalog.

Figure 13.5 Two alternative ways to create a list of function values.

Summing sequences to create left- and right-hand sums

The previous examples created simple lists. To sum such a list, we can use sum. The translation from mathematical symbols to TI commands is

$$\sum_{x=0}^{4} x^2 \Rightarrow \text{sum(seq(X^2,X,0,4,1))}$$

A common calculus task is to form the left- and right-hand sums for a function over an interval that has been divided into *n* subintervals. This is slightly more complicated than what we have done, but it is just a sum of terms formed by a function value times the length of an interval subdivision. Geometrically, it is the sum of areas of a collection of rectangles that are $f(x)$ high and Δx wide. That is, they approximate the area under the function's graph. In Chapter 16 we will use a program that draws the rectangles and calculates left- and right-hand sums, but in this chapter we just want to understand how to build these sums.

13. LEFT- AND RIGHT-HAND SUMS

The left- and right-hand sums: `sum(seq(...))`

We will break this task into three parts:

- Define `y1`, the function whose values we sum, and `y2` and `y3` with the left- and right-hand sum formulas
- Set the interval and number of subdivisions, and calculate the subdivision length
- Evaluate the sums as `y2` and `y3`

There is a bit of a trick here. By storing the sum formulas in the function definition of `y2` and `y3`, we can preset all the variables used in these formulas and then the sum will be the value of the function. The functions `y2` and `y3` are defined in Figure 13.6 with the following general format

$$\texttt{sum(seq(} f(x_{index})*\Delta x, \text{ index, start, stop, increment}\texttt{))}$$

These functions have no x variable, so they are constants. The left-hand sum, `y2`, starts at the left of the first interval so its x-values are A, $A + D$, $A + 2D$, etc., where A is the left endpoint and D is the length of the subinterval. For the right-hand sum, `y3`, we start at $A + D$ (with $I = 1$) and go until $I = N$, where N is the number of divisions. First enter the function `y2` and then use `2nd_ENTRY` to edit the `y2` definition into `y3`. (TI-85 users can replace `y1(A+I*D)` with `evalF(y1,x,A+I*D)` in the `y2` and `y3` formulas shown.)

The next step is to define the variables in the sum formula. Recall that the colon allows us to enter several commands at once. The `.1` shown is the result of the last command; it is the value of `D`. This provides a reality check on the values you entered. To repeat find a sum with a greater `N`, just use `2nd_ENTRY` to repeat the previous commands and edit the long command line.

```
y1=1/x
                    Done
y2=sum (seq(y1(A+I*D)
,I,0,N-1,1)*D)
                    Done
y3=sum (seq(y1(A+I*D)
,I,1,N,1)*D)
```

```
1→A:2→B:10→N:(B-A)/N→
D
                    .1
y2
           .718771403175
y3
           .668771403175
```

Figure 13.6 The left- and right-hand sums of $f(x) = 1/x$ over $1 \leq x \leq 2$ with 10 subdivisions. (TI-85 uses an alternate form of `y2` and `y3`.)

The final step is easy just request the value of `y2` and `y3`.

▶ *Tip: This is a place where* `2nd_INS` *and* `2nd_ENTRY` *are extremely handy. Also note that the* `STO→` *key puts* `ALPHA`*-lock on, so you must press* `ALPHA` *to unlock this mode and enter the numbers.*

► *Tip: You can store the two sum formulas in y98 and y99 to keep them out of the way of the usual functions and not write over them. Deselect them so that they do not effect later graphing.*

Negative values in the sum

Next we use the left-hand sum for the function $f(x) = \sin(x^2)$, first on the interval $0 \leq x \leq \sqrt{2\pi}$ and then on the shorter interval $0 \leq x \leq \sqrt{\pi}$. The graph in Figure 13.7 shows that the function is negative from $\sqrt{\pi}$ to $\sqrt{2\pi}$. The fact that that the sum on the longer interval $0 \leq x \leq \sqrt{2\pi}$ is less than the sum on the subinterval $0 \leq x \leq \sqrt{\pi}$ is understandable since we are adding the negative values on the subinterval $\sqrt{\pi} \leq x \leq \sqrt{2\pi}$. This shows that we will need to be careful when interpreting these sums as areas.

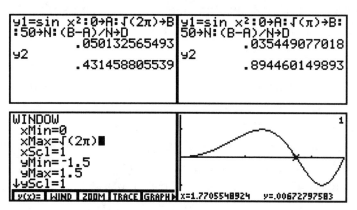

Figure 13.7 *The total sum over an interval may be less than the sum over a subinterval if the function is negative.*

Approximating area using the left- and right-hand sums

By increasing the number of partitions, the left- and right-hand sums may approach a limit which we interpret as the (signed) area under the function's graph. We write this as

$$\int_a^b f(x)dx = \lim_{n \to \infty} \sum_{i=1}^{n} f(x_i) \Delta x$$

As an example, let's see if there is a limit to the left hand sums of $\sin(x)$ over the interval $0 \leq x \leq \pi$ as the number of partitions increases. The computations with $N = 10$, 25 and 50. in Figure 13.8 suggest that the sums approach 2. We will later confirm that $\int_0^{\pi} \sin(x)dx = 2$ by using the Fundamental Theorem of Calculus.

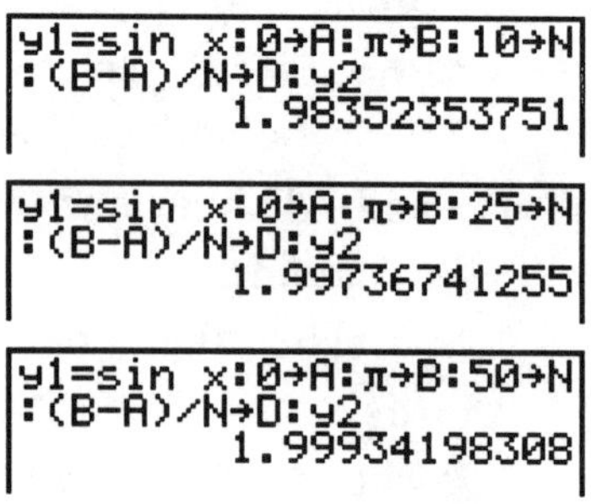

Figure 13.8 *The left-hand sum of the sine function on the interval $0 \leq x \leq \pi$ approaches the value 2.*

14. THE DEFINITE INTEGRAL

We have just seen in the previous chapter that we can calculate left- and right-hand sums which approximate the signed area under a curve. The definite integral is defined as the limit of the left-hand (or right-hand) sum as the number of partitions goes to infinity. Thus, each definite integral is a specific real number, and the TI will calculate this value. (Well, almost — it calculates an approximation that is generally reliable.) The definite integral is evaluated as a number, but we will see that we can also define its upper limit as a variable and thus create a new function.

The definite integral from a graph: ∫f(x)

After displaying a graph we can find the definite integral of a function *and* see its graphic representation. Let's start with $y = 2\sin(x)$ and graph it in the ZTRIG window. Here we are again in need of

GRAPH MORE MATH

to use the integral option ∫f(x) (F3). You are prompted to set the lower limit and then the upper limit. These limits are set in the same way that you have already set bounds using FMAX. Remember that they must be within xMax and xMin.

The number ∫f(x) can be interpreted as a measure of the shaded signed area under the function's graph. It may surprise you that the result is an integer, but notice that y1 is twice the sine function and recall from the last chapter that the left- and right-hand sums of the sine function converge to 2 over this interval.

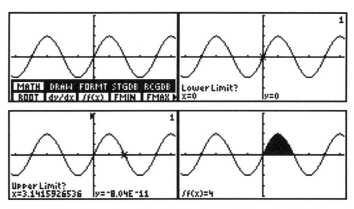

Figure 14.1 The value of a definite integral found and shown as shaded area on a graph.

➤ *Tip: If there is more than one function graphed, then you must select the desired function before using the ∫f(x) command.*

The definite integral as a number: `fnInt(...)`

To find the same result without using a graph, you can work from the home screen. The TI command for the definite integral is in the `2nd_CALC` menu as `fnInt` (F5). Its name comes from 'function integral' and it has the general format

$$\text{fnInt}(\textit{function, variable, lower, upper})$$

The translation from the mathematical symbols of the previous example to TI commands is

$$\int_0^\pi 2\sin(x)\,dx \Rightarrow \text{fnInt}(2\sin(x), x, 0, \pi)$$

Facts about the definite integral

Four definite integral facts will be illustrated below using simple functions and their graphs. You are encouraged to change the function and window to make it more exciting. In each of the following examples, we show the result graphically and then, in the final frame, the numerical rendition of the same result on the home screen. It is important that you feel comfortable using both methods of finding the definite integral.

After a graph with shading has been displayed, it is usually desirable to clear the screen before the next graphing. This can be done in a variety of ways.

- Press `GRAPH MORE DRAW MORE MORE CLDRW`. But this is hard to remember.
- Use a `ZOOM` menu selection. This is handy when your window comes from `ZSTD`, `ZPREV`, `ZTRIG`, `ZDECM` or `ZRCL`.
- Press `y(x)=` to reach the function definition screen, retype over the first (or any other) character. Then press `2nd_GRAPH` (F5). Even if formulas remain the same, edits force a fresh graph.

▶ *Tip: In order to get a fresh graph, you must edit a currently defined function in the `y(x)=` list; adding a new function will not cause any current functions to be graphed again.*

Reversed limit integrals are the negative of one another

Unlike most settings, where error messages are given whenever `xMin` > `xMax` or `Left Bound` > `Right Bound`, the `Upper Limit` and `Lower Limit` can be in either order. The result of an order reversal is a sign change of the value. See Figure 14.2.

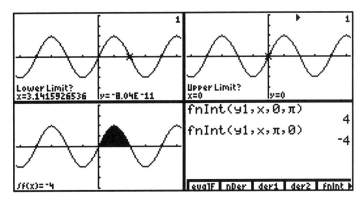

Figure 14.2 Limit reversal changes the sign of the definite integral.

The intermediate stop-over privilege

The definite integral can be calculated as a whole from the lower to upper limit, or it can be calculated in contiguous pieces. This can be thought of as a plane fare where the charge is the same whether you fly non-stop or have an intermediate landing. Figure 14.3 shows that we get the same answer dividing our function over two particular subintervals.

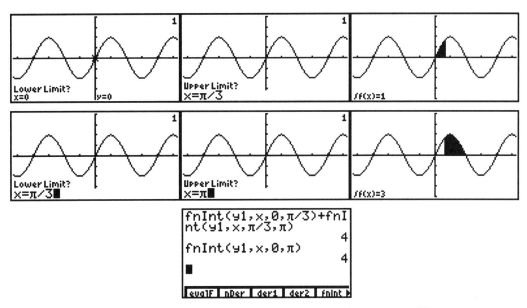

Figure 14.3 The definite integral found in two pieces. (Note: direct entry of limits is not available on the TI-85.)

The definite integral of a sum is the sum of the integrals

In Figure 14.4, we set y1= 2 sin x and y2 = x. The graph of the function y3=y1+y2 is shown as a bold curve. We see that the definite integral of the sum function, y3, is the sum of the definite integrals of the two summand functions.

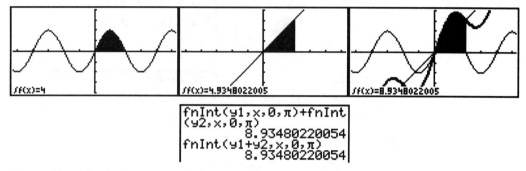

Figure 14.4 *The definite integral of a sum is the sum of the integrals. When computing the definite integral of* y3, *you will need to use the up arrow until a* 3 *appears in the upper right corner to have the TI refer to that function.*

Constant multiples can be factored out of a definite integral

We already saw an example of this with $\int_0^\pi 2\sin(x)dx = 2\int_0^\pi \sin(x)dx$. Another example is shown in Figure 14.5. We first check numerically to see that they are the same. Then we find $\int_0^\pi 3\sin x\,dx = 6$ from a graph.

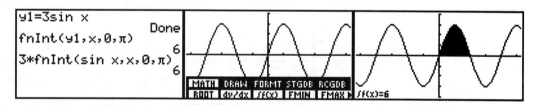

Figure 14.5 *A constant multiple of a function can be factored out of a definite integral.*

The definite integral as a function: y= fnInt(...,x)

Recall that der1(T², T, 1) is the (first) derivative value of the square function at 1 and, by replacing 1 with x, we can create a derivative function, der1(T², T, x). We repeat this dummy variable technique to create an integral function. We will use T as the dummy variable in the examples (it could be called x and it would make no difference). For example,

$$\int_a^x \cos(t)dt \Rightarrow \text{fnInt(cos T, T, A, x)}$$

14. THE DEFINITE INTEGRAL 85

➤ *Tip: Graphing functions defined with* `fnInt()` *is quite slow; setting* `xRes` *to a higher number will increase graphing speed.*

Figure 14.6 The integral function of the cosine appears to be the sine function. (Recall that the TI-85 has no table feature.)

In Figure 14.6, we choose 0 as the lower limit, so
$$y1 = \text{fnInt}(\cos T, T, 0, x),$$
and we graph it using a `ZTRIG` setting (and `xRes = 3`). We see that the graph of `y1` looks like a sine function and we check this using a table with `TblStart = 0` and `△Tbl = π/12`. (You could also graph the sine function and check that the two functions have the same graph.)

When we change the lower limit in the `fnInt()` definition, interesting things happen. You can see from Figure 14.7 that the three functions defined this way are vertical shifts of one another. Thinking of the derivative as the rate of change of *y*-values, these three should have the same derivative. Each of these three functions is called an antiderivative of cos(*x*). The common notation used is

$$\int \cos(x)\,dx = \sin(x) + C$$

where *C* is an arbitrary constant. Neither the TI-85 nor 86 calculator can give you symbolic solutions of this type, but the TI-92 does have this power.

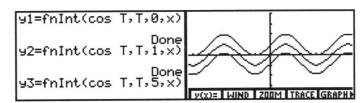

Figure 14.7 Antiderivatives of the cosine function are of the form sin(x) + C.

15. THE FUNDAMENTAL THEOREM OF CALCULUS

The Fundamental Theorem of Calculus will be discussed in two forms: as the total change of the antiderivative, then as a connection between integration and differentiation.

Why do we use the Fundamental Theorem?

The Fundamental Theorem of Calculus states

$$\text{if } f \text{ is a continuous function and } f(t) = \frac{dF(t)}{dt}, \text{ then } \int_a^b f(t)dt = F(b) - F(a)$$

One reason we use this theorem is that it calculates the definite integral in a simple way. Unfortunately, this use is limited to when we know the antiderivative. For example, we previously used the Riemann sum to guess that the definite integral `fnInt(sin x,x,0,π)` = 2. Using the Fundamental Theorem, we could do this calculation in our head:

an antiderivative of $f(x) = \sin(x)$ is $F(x) = -\cos(x)$,

so $F(\pi) - F(0) = -\cos(\pi) - (-\cos(0)) = -(-1) - (-1) = 2$

A second reason to use the Fundamental Theorem is that it gives us an exact answer, which may be required or just plain useful. For example, when a growth factor compounds continuously, the decimal accuracy is limited to that of the calculator. This is fine when we are dealing in thousands or millions, but sometimes we have amounts that are astronomical and we want an answer that will be exact to whatever number of decimal places are required. Think of the value of π: it is roughly 3, or, if more accuracy is needed, we can use 22/7, or better yet, 3.1459. In its exact form, the symbol π represents full accuracy, not a decimal or fraction approximation.

The definite integral as the total change of an antiderivative

Let's look at an example where an exact answer will be found. Consider a savings account into which you put a dollar every hour. What will it be worth in 20 years if it is compounded continuously at a 10% annual rate? This is a thinly disguised definite integral. First, whatever you deposit needs to be expressed in an annual amount so that all our rates are annual. Call this amount P, which we will take as 365*24 (ignoring leap years). Deposits are so frequent that we will consider the rate to be continuous. The future balance in ten years is then given by the definite integral

$$\int_0^{20} Pe^{0.1(20-x)}dx$$

The `fnInt()` function gives a value of over half a million dollars. This is probably good enough in this case, but the calculator answer does have limited accuracy. We now use the Fundamental Theorem to write an exact answer (correct to an infinite number of digits). This is sometimes called *closed form*. We first find an antiderivative function,

$$F(x) = Pe^2\left(-\frac{e^{-0.1x}}{0.1}\right)$$

and then evaluate it at the upper and lower limits:

$$F(20) - F(0) = Pe^2\left(-\frac{e^{-2}}{0.1}\right) - Pe^2\left(-\frac{e^{-0}}{0.1}\right) = P\left(\frac{e^2}{0.1}\right)\left(1 - e^{-2}\right)$$

When written in terms of *e*, the expression has full accuracy. The two answers are the same when compared with the accuracy shown in Figure 15.1. Knowing the closed form solution allows us to accurately find the future value of saving a thousand dollars per minute, although we would need a calculator with more internal digits of accuracy to see the difference.

```
fnInt(365*24 e^(.1(20
-x)),x,0,20)
             559681.314266
365*24((e^2)/.1)(1-e^
-2)
             559681.314266
```

Figure 15.1 A definite integral calculated using the built-in numeric integral approximator and the closed form given by the Fundamental Theorem.

A note to TI-85 users on function evaluation notation

The standard notation in the Fundamental Theorem is *F(b) – F(a)*, which we cannot mimic on the TI-85. However, you can accomplish the same calculations by using the `evalF()` command. For example, using `y1=x²`, the following are equivalent on the two calculators:

TI-86: `y1(2)-y1(0) = 4`

TI-85: `evalF(y1,x,2)-evalF(y1,x,0) = 4`

It should be clear that the TI-86 notation is superior. From now on, TI-85 users are expected to convert to `evalF` when function evaluation notation appears.

Using fnInt() to check on the Fundamental Theorem

Consider the example of finding the value of $\int_0^2 x^2 dx$ in different ways. First let's graph the function $y = x^2$ and use **2nd_CALC** $\int f(x)$. This is shown in Figure 15.2. The last frame is on the home screen, where we first use **fnInt()** and then the Fundamental Theorem with $F(x) = x^3/3$ to evaluate the integral numerically as $F(2) - F(0)$. Luckily we got the same answer all three ways.

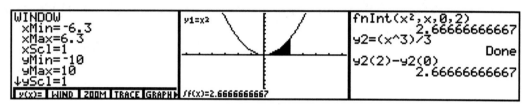

Figure 15.2 *Different aspects of the Fundamental Theorem equation.*

▶ *Tip: If you have trouble getting to a limit that you want by using the arrow keys, even using ZDECM, then the xRes may need to be reset to 1.*

Viewing the Fundamental Theorem graphically

In the above examples, the lower and upper limits were constants. Now consider the upper limit as a variable. Specifically, let the upper limit be a variable on each side of the Fundamental Theorem equation, which gives a function in terms of x:

$$\int_a^x f(t)dt = F(x) - F(a)$$

We now try an example by setting up the following functions:

$f(x) = y1 = .1x^2$, whose antiderivative is $F(x) = y3 = (.1x^3)/3$

Since a is arbitrary, let $a = -3.5$. (This means that our summing starts at zero when $x = -3.5$.) Finally, we define the two functions we want to compare and see if they are equal:

$$y2 = fnInt(y1, x, -3.5, x) \text{ and } y4 = y3(x) - y3(-3.5)$$

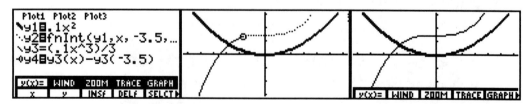

Figure 15.3 *An example of $\int_a^x f(t)dt = F(x) - F(a)$ with $f(t) = 0.1t^2$.*

15. THE FUNDAMENTAL THEOREM OF CALCULUS

Using a ZDECM graphing window (with xRes=2) and the graphing styles shown in the first frame of Figure 15.3, you will see that y2 and y4 have the same graph.

Checking on f(x) = e^x, the function that is its own antiderivative

As another example of this kind of comparison, we let $f(x)$ be the famous exponential function, whose antiderivative is itself. We check that the two sides of the Fundamental Theorem equation are graphically equal by defining and graphing the two functions shown in Figure 15.4.

➤ *Tip: If you are verifying that two functions have the same graph, use TRACE and the up arrow to move back and forth between functions.*

➤ *Tip: Graphing with fnInt as part of a formula is very slow. Patience is required, especially for TI-85 users. TI-86 users can speed it up some by increasing the xRes setting.*

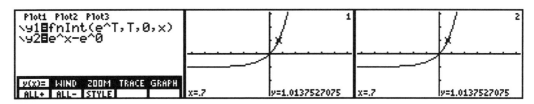

Figure 15.4 An example of $\int_a^x f(t)dt = F(x) - F(a)$ with $f(t) = e^t$.

Comparing nDer(fnInt(...)...) and fnInt(nDer(...)...)

What happens when you find the derivative of the integral function? It should not be too surprising that you get the original function back. Consider $g(x) = \sin(x)/x$, whose domain is all non-zero real numbers.

Here is another example where der1 will give an error message, so we must use the numeric derivative nDer. In the previous example we used the TRACE to compare values. In this example we show how STYLE can distinguish functions. The y1 has dot style and y2 has a leading cursor style. We use a ZDECM window setting.

In both the graph and table of Figure 15.5, it looks like y1 and y2 match perfectly, except at $x = 0$. However, there is so much numerical action taking place in the calculation of y1 that these values are just close approximations. This can be seen in the table if you highlight a y1 entry and compare it to the corresponding y2 value: they are not exactly the same.

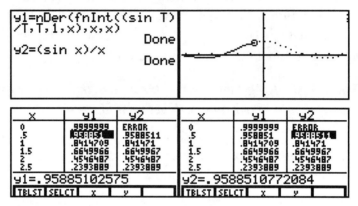

Figure 15.5 Graph and table for nDer(fnInt(...)...) *with closer inspection of two table entries that should be equal but differ slightly because of approximation errors.*

Let's reverse the order and integrate the derivative function. You might think you will get the original function back again but, as you can see in Figure 15.6, you get the original function plus some constant (which depends on the lower limit you enter).

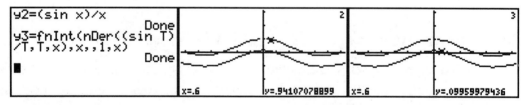

Figure 15.6 Graph showing that fnInt(nDer(...)...) *differs from the original function by a vertical shift of -sin(1) = -0.84147...*

16. RIEMANN SUMS

In Chapter 13 we introduced the left- and right-hand sums to approximate the definite integral. In this chapter we use a program to simplify explorations of these and other types of sums. We will add the capability to graphically view the subdivision areas that sum to make the approximation.

A few words about programs

This marks our first use of a stored program, so perhaps an introduction is in order. A program is a set of commands that are performed in a prescribed order, like a recipe. The order is normally the sequential list of commands, but there are techniques to alter that order. Special program command menus (**I/O** and **CTL**) list the input/output and control commands.

A program is written by pressing **PRGM EDIT (F2)** and entering a new name. If you enter the name of an existing program, then that program will be put on the screen for editing. For a new program, enter the desired sequence of commands. Commands are entered by pasting from menus or the catalog; they can also be typed directly, but you need to pay special attention to upper and lower case letters (the calculator is case sensitive). A colon starts each new program line. Use **EXIT EXIT** to exit the program editing mode.

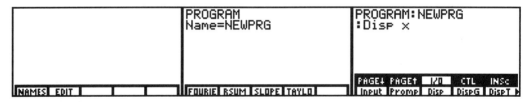

Figure 16.1 Start with a fresh home screen and chose **EDIT**. *Enter a new program name and then a list of commands.*

A program is activated — the more common terms are *run* or *executed* — by pressing **PRGM NAMES (F1)**, selecting the program name, and then pressing **ENTER**. Step-by-step instructions can be found in the programming chapter of the *Guidebook*.

Entering a program from the printed page is quite tedious and you can commonly expect to make a few errors that will only be discovered when executing the program. However, once a program is correctly entered in a TI calculator, it can be transferred to other TI calculators using the built-in **2nd_LINK** commands. In the classroom, it is typical that a program is verified by the instructor and distributed to the class using **2nd_LINK**. You can download additional programs from the Web; this is outlined in the Appendix.

The program pasting error

The most common error when you first use programs is to forget that programs are pasted onto the home screen and then executed. For example, you might be in the middle of creating a command to sum a sequence when you remember that you have a program to do it. Pressing PRGM and selecting the program name RSUM will paste the command RSUM at the end of the line you were on. Some kind of error message results because the PRGM command was not on its own line.

Figure 16.2 Pasting a command to execute a program can cause an error if the home screen cursor is not on a new line.

Using the RSUM program to find Riemann sums

A program to automate the Riemann sum process is given at the end of this chapter. It is designed to require a minimum of input to the program. Before using it, you must define your function in y1 and set the window to have xMin be the left endpoint and xMax be the right endpoint of the desired interval. In Figure 16.3 we take these two preliminary steps.

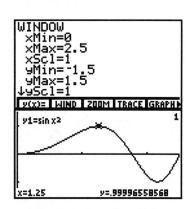

Figure 16.3 Set y1 and the window before using the RSUM program.

In Figure 16.4 we first see the program name RSUM pasted from the submenu and started from the home screen. A few moments after activation, the program prompts you for the number of subdivisions (partitions) of the interval. Next, five different kinds of Riemann sums are calculated and displayed. The value labeled fnInt is short for fnInt(y1,x,xMin,xMax) and can be used to judge the other numerical approximations. The last value, labeled Simpson, is the value given by Simpson's method. This is a weighted average of the two previous results, specifically the sum of the trapezoid value and twice the midpoint value all divided by three. You will find that this value is consistently close to the 'true' value for small partitions.

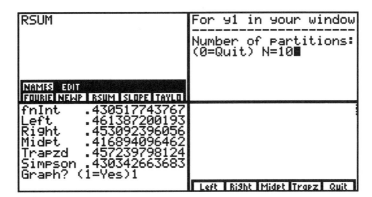

Figure 16.4 The RSUM program gives six numeric approximations of the definite integral of y1 from xMin to xMax.

In Figure 16.5, the graphic representations of the Riemann sums are shown in the order of the four menu choices on the menu line. A set of vertical dots shown in the upper right corner will appear on the active screen as a twinkling. This alerts you that the program is in Pause mode; it is ready for you to make the next choice from the menu line. Pressing Quit will allow you to reset the number of partitions.

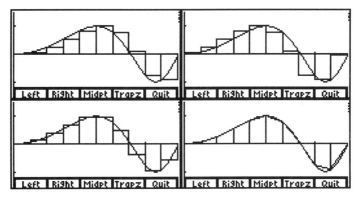

Figure 16.5 The four graphic representations of the Riemann sum: left, right, midpoint and trapezoid.

The TI-86 and TI-85 program RSUM

The program listed below takes a considerable amount of time to enter, but once entered, can be passed to other TI calculators of the same type. (It is in the public domain.) It can also be stored on a computer using the TI Graph Link.

For faster entry, you can use a copy technique on repetitive code. The Lbl A1 through Lbl A5 sections are similar. You can enter one section as a new program and repeatedly use **2nd_RCL PRGM** *progname* to paste in four copies of the section, then edit them in the few places where they differ.

```
Lbl N1
ClLCD
Disp "For y1 in your window"
Disp "--------------------"
Disp "Number of partitions:"
Input "(0=Quit) N=",N
If N<1
Stop

ClLCD
(xMax-xMin)/N→D
Disp "fnInt"
fnInt(y1,x,xMin,xMax)→U
Outpt(1,9,U)
Disp "Left"
sum(seq(evalF(y1,x,xMin+I*D),I,0,N-1,1)*D→L
Outpt(2,9,L)
Disp "Right"
sum(seq(evalF(y1,x,xMin+I*D),I,1,N,1)*D→R
Outpt(3,9,R)
Disp "Midpt"
sum(seq(evalF(y1,x,xMin+I*D),I,.5,N,1)*D→M
Outpt(4,9,M)
Disp "Trapzd"
(L+R)/2→T
Outpt(5,9,T)
Disp "Simpson"
(2M+T)/3→S
Outpt(6,9,S)

Input "Graph? (1=Yes)",A
If A≠1
Goto N1
ClLCD
Func
FnOff
FnOn 1

Lbl G1
Menu(1,"Left",A1,2,"Right",A2,3,"Midpt",A3,4,"Trapz",A4,5,"Quit",A5)

Lbl A1
ClDrw
For(I,0,N-1)
xMin+I*D→X
X→x:y1→Y
Line(X,0,X,Y)
Line(X,Y,X+D,Y)
Line(X+D,Y,X+D,0)
End
Goto G1
```

```
Lbl A2
ClDrw
For(I,0,N-1)
xMin+I*D→X
X+D→x:y1→Y
Line(X,0,X,Y)
Line(X,Y,X+D,Y)
Line(X+D,Y,X+D,0)
End
Goto G1

Lbl A3
ClDrw
For(I,0,N-1)
xMin+I*D→X
X+D/2→x:y1→Y
Line(X,0,X,Y)
Line(X,Y,X+D,Y)
Line(X+D,Y,X+D,0)
End
Goto G1

Lbl A4
ClDrw
For(I,0,N-1)
xMin+I*D→X
X→x:y1→Y
Line(X,0,X,Y)
X+D→x:Line(X,Y,X+D,y1)
End
xMax→x:Line(xMax,y1,xMax,0)
Goto G1

Lbl A5
Goto N1
```

17. IMPROPER INTEGRALS

In this chapter we will look at two different but similar problems encountered when trying to use the integral in a wider setting. These special cases that involve infinity are called improper integrals. First, we will see that sometimes the limit of integration can be infinite. Second, we will see that integration is sometimes possible even when the integrand function itself has infinite values.

An infinite limit of integration

On a first take, you might think that any positive function that goes on forever must have an infinite definite integral. Let's try a thought experiment. Suppose you decided to go on a diet and every day you cut your chocolate chip cookie consumption in half. How many cookies would you need for your lifetime? (or for eternity?) Visualize the cookie: the first day you would eat half, the next day a half of a half (a quarter), and so on. Because each day you would only eat half of the remaining cookie, you would never finish it: one cookie would last a lifetime! A graph for ten days can be drawn using the tricky function `y1 = (1/2)^int(x+1)`. The `int()` function is found in `2nd_MATH NUM` (above `F4`) and gives the integer part of a number, so `y1` is $\frac{1}{2}$ until 1, then $\frac{1}{4}$ until 2, etc. Drawing the graph in the standard connected style gives a slight vertical jag getting from $\frac{1}{2}$ to $\frac{1}{4}$, etc., due to the sampling problem that we have seen and explained before. Figure 17.1 shows the graph drawn with the dot style, along with the `∫f(x)` computation that suggests the total area under the graph is 1.

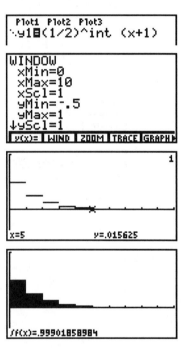

Figure 17.1 The sum of cookie halving is one.

Three power functions

Let's compare the three functions

$$y_1 = \frac{1}{x}, \quad y_2 = \frac{1}{x^3}, \quad \text{and} \quad y_3 = \frac{1}{x^{1/3}}$$

as x goes from 1 to infinity and see if we get a graphical and numerical hint about which might have a finite sum. To have a finite sum, a continuous function must approach zero. We see in Figure 17.2 that all three functions do approach zero as

17. IMPROPER INTEGRALS   97

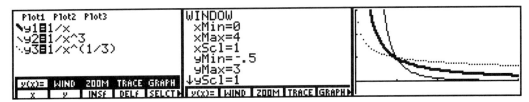

Figure 17.2 Comparing three functions as x goes to infinity. y2 *goes to zero the fastest.*

x gets large, and y2 does so the fastest. In Figure 17.3 we try to see if $\int_{1}^{\infty} \frac{1}{x} dx$ exists by finding $\int_{1}^{10} \frac{1}{x} dx$ and $\int_{1}^{100} \frac{1}{x} dx$. We see that the integral as x approaches infinity does not seem to be converging to any value — at least not right away.

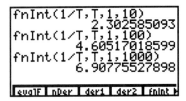

Figure 17.3 The integral values as x gets larger do not appear to converge.

➤ *Tip: When using the* fnInt() *function, expect long calculation times. If the wait is too long to bear, press* ON *to halt the computation.*

Graphing the integral with the upper limit as variable

A different approach is to graph the values of the definite integral as x becomes large. Unfortunately, this takes several minutes. You can set it up to graph and take a break while it works. We see that the graphs of y1 and y3 appear to be ever increasing without bound, but the graph of y2 appears as a horizontal line at $y = 0.5$. Who knows, maybe y1 and y3 become constant for very large x. Actually, we do know that the first integral is defined to be $\ln(x)$, so it does not converge to any finite number. Notice in Figure 17.4 that the graph of y1 looks like the graph of $\ln(x)$.

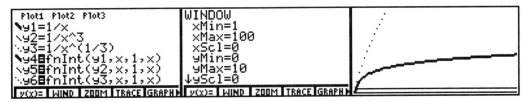

Figure 17.4 Graphs of the integral functions as the upper limit gets large. The horizontal line at the bottom of the screen suggests that the graph of y2 *converges.*

Making a table of integral values as the upper limit increases

The table mode works well for checking the values of a definite integral as *x* increases. In our previous encounters with the table, we used a start and increment. For our current use, it is better to set **Indpnt:** to the **Ask** setting. After doing this in **TBLSET**, the **TABLE** command will give you a blank column of *x*-values. We fill this column with larger and larger values (we have freedom to choose each *x*-value). Each row will take longer to calculate. You can use the right arrow to reach **y6** and find its values, but shifting to the right (as shown in Figure 17.5) means that all the shown values must be calculated, so it is extremely slow.

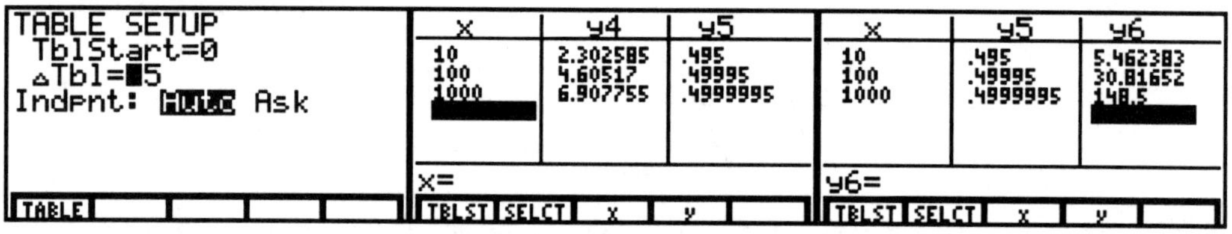

Figure 17.5 Table values of the integral functions as the upper limit gets larger.

A command line for integral evaluation

If you want to minimize keystrokes, you can enter the following simple three-piece command that calculates values for five integrals from $\int_{1}^{\infty} f(x)\,dx$ to $\int_{1}^{100000} f(x)\,dx$ as in Figure 17.5. Define the function in **y1** and enter the following sequence on the home screen.

```
For(I,1,5):Disp fnInt(y1,x,1,10^I):End
```

The truth about $\int_{1}^{\infty} \frac{1}{x^p}\,dx$

You can investigate other examples such as $Y = \frac{1}{x^2}$ and $Y = \frac{1}{x^{3/4}}$ to test the following analytic results.

$$\text{If } p < 1 \text{ then } \int_{1}^{\infty} \frac{1}{x^p}\,dx \text{ converges}$$

$$\text{If } p \geq 1 \text{ then } \int_{1}^{\infty} \frac{1}{x}\,dx \text{ diverges}$$

The convergence of $\int_0^\infty \frac{1}{e^{ax}} dx$, $(a > 0)$

It is simple to show analytically that $\int_0^\infty \frac{1}{e^{ax}} dx = \frac{1}{a}$ for $a > 0$. Let's try this out using the table values. In Figure 17.6 you see that the integral is already approximated to be 0.2 with an upper limit of 10, so it should continue to calculate that value for higher values, right? Indeed it should, but you can see, for larger values of the upper limit, that the approximation becomes zero! A moment's thought will tell you why it is wrong: the dreaded sampling problem lurks behind this error. As the interval becomes larger, the calculator no longer samples near the 0 and thus misses the values of the function that contribute to give 0.2.

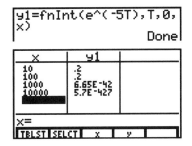

Figure 17.6 Table values of the integral functions as the upper limit gets larger and the accuracy deteriorates.

The integrand goes infinite

The second major problem that makes an integral improper is an infinite integrand (one version of being undefined). This is a much more dangerous situation because there is no ∞ symbol to alert you. It is a good habit to graph the function before finding the definite integral. The graph should alert you to potential problems, like the integrand being undefined (and tending to positive or negative infinity). Sometimes the calculator will warn you that it cannot approximate the integral; in such cases, you should suspect that the integral does not converge. We show this in Figure 17.7. Here we try to blindly find the definite integral with the value of the integrand being infinite at $x = 2$. By drawing a graph, we can see that we must be careful. There are cases where we can find an improper integral across a point where the integrand goes to infinity. The second example on the next page is such a case.

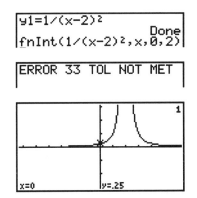

Figure 17.7 An error message for a divergent integral using fnInt() and a graph showing an infinite integrand.

Results we can trust

From the graph screen, if you use the `2nd_CALC` $\int f(x)$ option to evaluate an improper integral, then calculation time is extended and in some cases you will receive an error message. This could signal that the integral diverges, but you should be careful since this is not a reliable test. The example in Figure 17.8 shows a (correctly) failed attempt to evaluate a divergent integral.

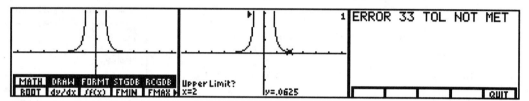

Figure 17.8 An error message when using $\int f(x)$ for the divergent integral `y1=1/x^4` for $-1 \leq x \leq 2$

Results we can't trust

The calculator is a very limited technological device and not completely reliable in handling improper integrals. The calculation time is long and the result is mildly undependable. The example in Figure 17.9

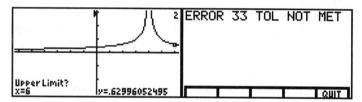

Figure 17.9 An error message when using $\int f(x)$ for the convergent improper integral `y2 = 1/(x-4)^(2/3)` for $0 \leq x \leq 6$.

gives an error message, suggesting that the integral diverges, when we know from the analytic results above that it converges. By using `2nd_MEM TOL` we can reset the tolerance from the default `tol=1E-5` to `tol=.01`. The calculator then finds that the integral converges. See Figure 17.10.

A calculus teacher can easily make up examples where the calculator will mislead you. Would someone really do that? Yes.

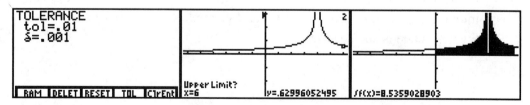

Figure 17.10 The convergent improper integral from Figure 17.9, evaluated with `tol=.01`.

18. APPLICATIONS OF THE INTEGRAL

We will look at four diverse applications of the integral. These standard examples give only a flavor of the extensive applications of this core mathematical concept.

Geometry: arc length

The following calculus formula finds arc length along a function's graph.

$$\text{Arc length} = L = \int_a^b \sqrt{1 + (f'(x))^2}\, dx$$

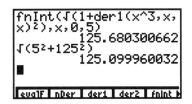

Figure 18.1 Finding arc length using an integral.

Let's calculate the arc length of the curve $y = x^3$ from $x = 0$ to $x = 5$ and compare it to direct distance from the origin to the point (5,125). In Figure 18.1, we see this result calculated on the home screen.

The TI-86 and TI-85 will calculate arc length as a command. We use **GRAPH MORE MATH MORE** to find the **DIST (F4)** and **ARC (F5)** keys. The usual Left and Right Bound prompts appear. The results are shown in Figure 18.2. The graph's scale makes the difference between the two lengths look greater than the actual numerical difference.

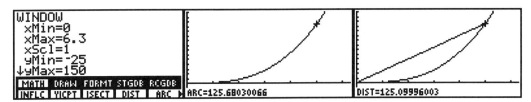

Figure 18.2 Finding the arc length of the curve $y = x^3$ from $x = 0$ to $x = 5$ and comparing it to the direct distance between (0,0) and (5,125).

Physics: force and pressure

Pressure increases with depth so that there is more pressure at the bottom of a container than at the top. One cubic foot of water weighs 62.4 pounds and it has a force of 62.4 pounds on the base. If it is only half full then it has a force of 62.4/2 = 31.2 pounds on the base. The pressure on the base is directly proportional to the depth of the water. We also know that force is the product of pressure and area.

The difficult part of pressure, force, or volume problems is not the actual integration that is required, but setting up the integral to accurately reflect the geometry of the situation. A time-honored system is to write the pressure as a sum of forces acting on strips or slices.

The pressure on a trough

Consider the trough shown in Figure 18.3. There are four sides that the force of water acts on. Let's tackle the easiest side first: the 3' by 14' horizontal back. We subdivide the height into pieces of length Δh, so this back side is made up of horizontal strips, each having an area of $14\Delta h$ square feet. The entirety of each strip is at the same depth, so that all along a given strip there is equal pressure, namely:

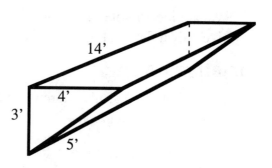

Figure 18.3 A trough with the shape of a triangular prism.

$$\text{force on a horizontal strip} = (62.4h)(14\Delta h)$$

(We could multiply the constants, but the calculator can do that and we choose to leave our setup in this more readable form.) The total force is sum of all the horizontal strips. The exact total is given by the definite integral

$$\int_0^3 (62.4h)(14)dh$$

Next we consider the inclined side. For a Δh (vertical height change) the actual width w on the inclined side is greater than Δh. Using similar triangles, these values are related by $\Delta h/w = 3/5$. The area of the strip is $14(5/3)\Delta h$, so

$$\text{force on inclined strip} = (62.4h)(14)(5/3)\Delta h$$

The total force is given by the integral

$$\int_0^3 (62.4h)(14)(\tfrac{5}{3})dh$$

Finally we compute the force on the two triangular ends. Again using similar triangles, we find the area of a strip is $(4/3)(3-h)\Delta h$ and

$$\text{force on end strip} = (62.4h)(4/3)(3-h)\Delta h$$

so that the third integral is

$$\int_0^3 (62.4h)(\tfrac{4}{3})(3-h)dh$$

18. APPLICATIONS OF THE INTEGRAL

Now the easy part is entering these definite integrals for evaluation and computing the final sum; this is done in Figure 18.4.

Figure 18.4 *The three definite integrals to find the force on a trough.*

➤ *Tip: It is much safer to leave calculator results in an expanded form so that the derivation remains evident.*

Economics: present and future value

Suppose you win a two million dollar lottery. Before you spend the money, you are told that the money will be distributed to you over the next twenty years. That is a mere $100,000 per year (before taxes). If you wanted to get an immediate lump sum, you could sell your rights to all the future payments. What is this really worth? In economics this value is called the present value, V.

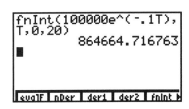

Figure 18.5 *The present value of a 2 million dollar lottery.*

It can be calculated using a fixed investment rate and an integral. If P is the annual payment and r is the annual investment rate for T years, then the present value is

$$V = \int_0^T Pe^{-rt}dt$$

Suppose that the agreed investment rate was 10% for the twenty years. Then we enter the integral as shown in Figure 18.5. The present value is $864,664.72, less than one million!

Using SOLVER for "what if" analysis with fnInt()

You can use fnInt() inside the SOLVER to set up an interactive session to both evaluate the present value and to see the effect of different investment rates. The rate is *r* and *x* is a dummy variable (which we set to 0 and essentially ignore). The SOLVER is very slow when there are integrals involved, so we speed up the process by using a wider tolerance (tol=.1) and still calculations take over one minute.

First we enter the equation

V=fnInt(Pe^(-r*x),x,0,T)

and check our setup by recalculating the present value of twenty annual payments of $100,000 when the investment rate is 10% ($r = 0.1$). We get the same answer as in Figure 18.5.

Figure 18.6 **SOLVER** *finding the present value V and then an investment rate r.*

Next we solve for an investment rate given a present value of $500,000. Remember that recalculation takes place when you place the cursor on the desired variable's line and press **SOLVE (F5)**. We find a present value of half a million corresponds to an investment rate of about 20% ($r = .196...$ in Figure 18.6).

▶ *Tip: Remember to change the* **tol** *setting back to its default (***tol=1E -5***).*

Discrete vs. continuous compounding

The analysis above makes the assumption that the income is coming continuously, effectively like every second, which is not the case. In real life, the money comes in twenty discrete payments. If we want to calculate this to the penny, we need to use a discrete sum for the twenty years. The point is that there is a close relationship between the integral and the sum of a sequence. The integral must be used when the income is continuous, but can be used as an approximation for a discrete

Figure 18.7 The present value is greater when discrete payments are made at the start of each year.

sequence sum. The sum, about $900,000 (see Figure 18.7), is more than the integral calculation because you receive all $100,000 at the start of the year. You can think of the continuous model as being paid about three-tenths of a cent per second for the next twenty years.

The future value $V = \int_0^T Pe^{r(T-t)}dt$

Since the probability of winning the lottery is essentially zero, you might want to create a jackpot for yourself by investing $100,000 a year (and remember, that's just three-tenths of a cent per second). Your total after T years is called the future value and is given by the heading's definite integral.

Using `fnInt()`, you can find that in twenty years you would have a real jackpot worth over six million dollars.

The integrals for present and future value are easily written in exact form using the Fundamental Theorem of Calculus. We find that

$$\text{future value} = (e^{-2}) * (\text{present value})$$

This is an example of the usefulness of the Fundamental Theorem: you would not be able to account for the e^{-2} factor without writing the integrals in symbolic form.

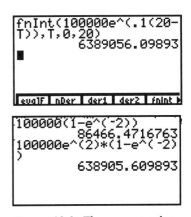

Figure 18.9 The present value and future value in exact form.

Modeling: normal distributions

In statistics, a normal distribution has a graph that is a bell-shaped curve. Its general equation is

$$p(x) = \frac{1}{S\sqrt{2\pi}} e^{-(x-M)^2/(2S^2)}$$

where M is the mean and S is the standard deviation. (In a statistics course, you would use the variables μ and σ for these values.) If $M = 0$ and $S = 1$, then the curve is called the *standard normal curve*, as shown in Figure 18.10 with $-3 \le x \le 3$ and $-0.1 \le y \le 0.5$. We saw an example of this family in the last section of Chapter 12.

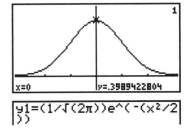

Figure 18.10 The standard normal curve.

The mean and standard deviation

To find the mean and standard deviation of any list, you can use `2nd_STAT CALC OneVa (F1)` and type the name of the list. The symbol for the mean is 'x-bar' ($\bar{x}$). There are two symbols for the standard deviation, a statistical indicator that tells how far values are spread. Without getting technical, use `Sx` if the list is a sample and `σx` for a complete list; as the number in the sample becomes large, there is almost no difference between these two.

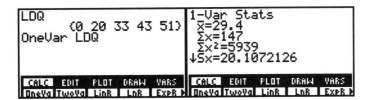

Figure 18.11 Finding the mean and standard deviation for a list.

The Anchorage annual rainfall

One application of the normal distribution is to model situations where measurements are taken under conditions of randomness. For example, suppose you look at the records for annual rainfall in Anchorage, Alaska over the past 100 years. Let's simplify and say that you found the average of these averages to be 15 inches. Let's say that the standard deviation was 1.

We can estimate the fraction of the years that rainfall is between

(a) 14 and 16 inches, (b) 13 and 17 inches, and (c) 12 and 18 inches

by taking three integrals of the normal distribution with $M = 15$ and $S = 1$. Using 2nd_CALC ∫f(x), we see from the graphs in Figure 18.11 that the model predicts

(a) 68% of the years have rainfall between 14 and 16 inches,
(b) 95% of the years have rainfall between 13 and 17 inches, and
(c) 99% of the years have rainfall between 12 and 18 inches.

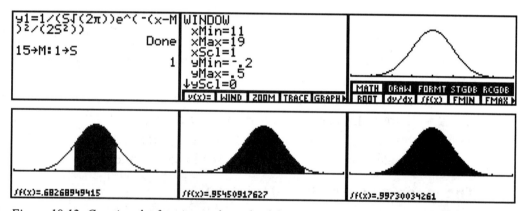

Figure 18.12 Creating the function and graph of the normal curve with mean 15 and standard deviation 1. Then the integrals with limits (a) $14 \leq x \leq 16$, (b) $13 \leq x \leq 17$ and (c) $12 \leq x \leq 18$.

▶ *Tip: After graphing an integral, use EXIT to return to the menu to set up the next graph. It is convenient to order nested intervals from the inside out, as in Figure 18.12. If a subsequent shading does not include the previous one, you will need to use DRAW MORE MORE CLDRW to clear the drawing.*

A note about TI statistical features

The TI-86 has powerful statistical features that are not addressed here. See Chapter 14 in the *TI-86 Guidebook*. (There are more limited statistical features on the TI-85, presented in Chapter 15 of the *TI-85 Guidebook*.)

PART IV
SERIES

19. TAYLOR SERIES AND SERIES CONVERGENCE

20. GEOMETRIC SERIES

21. FOURIER SERIES

19. TAYLOR SERIES AND SERIES CONVERGENCE

The use of series for approximation can be shown for simple cases on the TI-86, but a more powerful calculator or computer is needed for any serious work. On a more powerful calculator, such as the TI-92, a Taylor series command is built in. We will limit ourselves here to confirming a few well-known results. It is assumed that you know the formula for the Taylor polynomial.

The Taylor polynomial program

In our limited scope, we will derive Taylor polynomials of degree up to eight. This is because neither the TI-86 nor TI-85 can nest derivative definitions beyond the second. Since we need to evaluate the derivatives of higher orders, we will enter them into y(x) definitions by hand (which means that we have to find them algebraically!). We then use a simple but timesaving program that will calculate the $f^{(n)}$(Center) values needed to create the coefficients. The $f^{(n)}$(Center) values will be stored in LV (for 'list values'). The final stage is to store the polynomial function in y10 so that we can graph it and the original function for comparison. TI-86 users can set the y10 style to bold before executing the program to better see the comparison. Name the program TAYLOR.

Taylor polynomial program listing

```
ClLCD
Disp "Put function in y1."
Disp "Derivs. in y2, ..."
Disp "Set nice window."
Disp "------------------"
Disp "It lists function"
Disp "values in LV, puts"
Disp "polynomial in y10"
Input "Go On? (0=N,1=Y) ",N
If N<1
Stop

ClLCD
Input "Degree? (2-8) ",R
R+1→dimL LV
Input "Center? ",C

C→x
y1→LV(1)
y2→LV(2)
If R≥2
y3→LV(3)
If R≥3
y4→LV(4)
If R≥4
y5→LV(5)
If R≥5
y6→LV(6)
```

19. TAYLOR SERIES AND SERIES CONVERGENCE

```
If R≥6
y7→LV(7)
If R≥7
y8→LV(8)
If R≥8
y9→LV(9)
y10=sum (seq((LV(I+1)/I!)(x-C)^I,I,0,R,1))
Func
FnOff
FnOn 1,10
DispG
```

➤ *Tip: An optional line to graph the polynomial using bold style (for the TI-86 only) can be inserted just before last line:* `GrStl(10,2)`.

Setting up before using TAYLOR

Enter the function in y1, then y1' in y2, y1" in y3, etc. Deselect all the functions except y1 and find a nice window for y1. Leave some room above and below the function graph so the approximation graph can be seen. Check the list LV to be sure it doesn't contain data you want — it will be erased when we run the program.

We start by using the TAYLOR program for graphing the Taylor polynomial of degree five and center $x = 1$ for the function $y = \ln(x)$. See a typical setup in Figure 19.1.

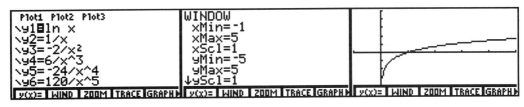

Figure 19.1 *Setting up to use the* TAYLOR *program.*

Following the prompts after pressing PRGM TAYLOR

After the setup, start the program from a fresh line on the home screen by pressing **PRGM NAMES** and **TAYLOR** from the **NAMES** submenu. This pastes the program name to the home screen, where pressing **ENTER** will start the program. After a short wait, the opening screen, as shown in Figure 19.2, is just a reminder that the y1 should be the original function and y2 should be the first derivative, y3 should be the second derivative, etc. You will want to preset the window so that it shows the function and has some extra screen space for graphing the Taylor polynomial, although you will be unsure exactly where the Taylor polynomial will

110 PART IV / SERIES

wander and may need to adjust your screen later. If you have done a setup, then press **1 ENTER** to continue.

▶ *Tip: In case you have not entered the function derivative definitions in the* y*'s and set the window or you have previously stored values in the list* LV *that you want to save, the* TAYLOR *program gives you a chance to exit (*0=N*).*

Next you will be prompted to enter the degree and center. The calculator will then compute and store the derivative values, $f^{(n)}$(Center), in the list LV. But note that these are not the actual polynomial coefficients until they are divided by $n!$, which is done in the function definition y10. The final prompt offers to graph your original function and the Taylor polynomial.

In Figure 19.2, we see the graphs of the function $y = \ln(x)$ and the fifth degree Taylor polynomial centered at 1 (in bold). The approximation looks very good when x is close to 1, but otherwise it is not very good, especially beyond $x = 2$.

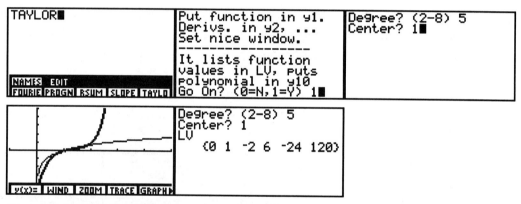

Figure 19.2 Typical screens for using the TAYLOR program. After graphing, the derivative values can be see in LV.

After TAYLOR has finished

As mentioned earlier, you can look at the values of $f^{(n)}$(Center) in LV. You can also change the degree variable R to any smaller degree value in the y10 definition and then graph again to see any lower degree Taylor polynomial.

The Taylor polynomials for $y = e^x$

Let's look at another example where knowing the higher order derivatives is easy. The function $y = e^x$ is its own derivative, so we can just repeat it in each yi. We now find the sixth degree Taylor polynomial with center 1. Previously, we used bold style to distinguish the Taylor polynomial on the graph. For variety, this

19. TAYLOR SERIES AND SERIES CONVERGENCE

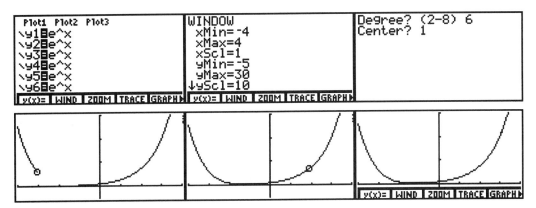

Figure 19.3 A Taylor polynomial of degree 6 centered at 1 will approximate $y = e^x$ when close to 1.

time we set the graph style of the Taylor polynomial in y10 to have a leading cursor in order to create a little animation. (TI-85 users cannot do this.)

In Figure 19.3 we see that the approximation is quite bad before $x = -1$ and then gives a good fit for the rest of the screen. We can only expect a polynomial to do a good job locally since $y = e^x$ approaches 0 as $x \to -\infty$ and no (nontrivial) polynomial does that.

Showing the interval of convergence for the Taylor series of the sine function

In the following example, we find the Taylor polynomials for $y = \sin(x)$ centered at 0. Enter the eight consecutive derivative functions and set the window with ZOOM ZTRIG. In Figure 19.4, only the graphs for degrees 1, 3, 5, 7 are shown because the graphs for degrees 2, 4, 6, 8 (respectively) are the same. This is because the even power Taylor coefficients are zero, as seen in LV. Looking at the four graphs, we can see that the approximations are

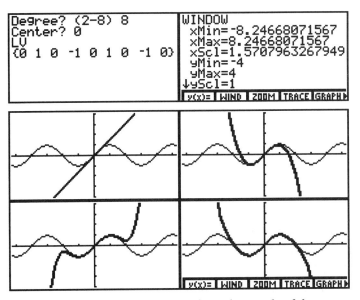

Figure 19.4 The graphs of the Taylor polynomials of degree 1,3,5,7 are shown in bold

close on wider and wider intervals as the degree gets larger. The interval of convergence for the seventh degree Taylor polynomial seems to be $-\pi < x < \pi$.

▶ *Tip: The higher the degree of the polynomial, the longer it takes to graph. TI-86 users may want to set xRes=4.*

The fact that we are approximating trigonometric and exponential/logarithmic functions is an example of the fact that any function that is infinitely differentiable can be locally approximated by polynomials. This is important because computers and calculators are masters at polynomial evaluation — after all, it is just addition and multiplication.

Evaluating a Taylor polynomial at $x = 1$ is the same as summing the coefficients. As the degree of the Taylor polynomial goes to infinity, the value at $x = 1$ remains finite because of the $n!$ term in the denominator of the Taylor coefficients. Another way of saying this is that the series of Taylor coefficients converges. This leads to a more general question.

How can we know if a series converges?

This is a difficult question, but we have the ratio test and the alternating series test to help us in many cases. If asked for proof, you will need to use an analytical argument, but a graph or table will often tell you what whether to pursue a proof of convergence or divergence. Since you are investigating a series, the nth term is given symbolically (or else you will need to write it symbolically). In the example below, the sigma summation notation is given, but if a series is listed without a symbolic term, then the first order of business is to write it as a sigma sum. This symbolic term is what we enter in the `seq()` command. We will use `y1 = sum(seq())` to either graph or make a table of values from which we form an opinion about whether the series converges or diverges.

The harmonic series: $1 + \dfrac{1}{2} + \dfrac{1}{3} + \dfrac{1}{4} + \ldots = \sum_{n=1}^{\infty} \dfrac{1}{n}$

We can translate the partial sums of the harmonic series into a TI function that sums the first through xth terms:

`y1=sum(seq(1/N,N,1,x,1))`

A good window for this kind of series investigation is xMin=0 and xMax=126: the *x*-values will all be sampled at integer values. By the 126th term, the long term behavior is usually evident. See Figure 19.5. The graphing gets progressively slower as *x* gets larger, but you can always press ON to break at any time.

19. TAYLOR SERIES AND SERIES CONVERGENCE

Table values are best found with `Indpnt:` set to `Ask`. Otherwise, it takes an inordinately long time to display one screen of table values. (You

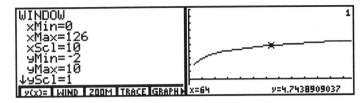

Figure 19.5 Suspecting divergence from a partial sum graph.

often get nothing but an error message because the calculator's capacity has been exceeded.) Figure 19.6 shows some table computations. You might ask why we don't just find the 10,000th partial sum. Without some previous values for comparison, it is unclear from one value whether the series is converging or diverging. On the practical side, you would have to wait several minutes for an answer, if you get one at all: attempting this in Figure 19.6 gives **ERROR 15 MEMORY**.

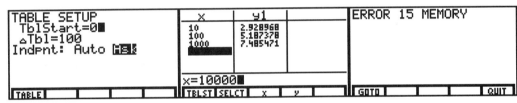

Figure 19.6 Suspecting divergence from a table, but getting an error message by going beyond the calculator's capacity.

The alternating harmonic series: $1 - \frac{1}{2} + \frac{1}{3} - \frac{1}{4} + \ldots = \sum_{n=1}^{\infty} (-1)^{n-1}(\frac{1}{n})$

Often series alternate positive and negative terms. This sign switch is cleverly written in sigma notation using a power of negative one. Since every other term is subtracted, we expect the range of the partial sums will not be as great as it is for the harmonic series.

▶ *Tip: When writing a sigma sum symbolically, it is easy to make small errors. Before graphing or making a table, you might want to make a small list of terms to verify that your notation is correct.*

We check the sequence
```
seq((-1)^(N+1)/N,N,1,5,1)
```
and then define the partial sums function
```
y1 = sum (seq((-1)^(N+1)/N,N,1,x,1))
```

Use $0 \leq x \leq 126$ so the calculations will be at integer values and $0 \leq y \leq 1$. Watch the graph and press **ON** to break when you are satisfied. We see from the graph in Figure 19.7 that it is a good bet that this series converges.

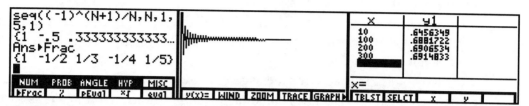

Figure 19.7 *Suspecting a convergence, we break before the graph is completed. We can also see evidence for convergence in the table values. (*MATH MISC ▶FRAC *helps us recognize the sequence entries.)*

A fast converging series: $2 - \frac{2}{3} + \frac{2}{9} - \frac{2}{27} + \ldots = \sum_{n=0}^{\infty} (-1)^n (\frac{2}{3^n})$

This series is related to the exponential function $f(x) = \frac{2}{3^x}$ and we know its values get close to zero as x gets large. In the first frame of Figure 19.8, we began with two mistakes: forgetting to start the sequence at zero, so that first term was not 2, and not enclosing the negative sign in an exponentiation. The corrected formula was used in y1.

▶ *Tip: Test your symbolic expressions on the home screen for proper use of parentheses, negative/subtraction signs, and syntax. Then use* **2nd_ENTRY** *to paste them in a* y *definition.*

Figure 19.8 *Detecting errors in notation then making a graph and table of a fast converging series.*

A slow converging series: $1 - \frac{1}{3} + \frac{1}{5} - \frac{1}{7} + \ldots = \sum_{n=1}^{\infty} (-1)^{n-1} \left(\frac{1}{2n-1} \right)$

Unlike the last example, some series converge very slowly. You may feel quite confident that a series is convergent, but find getting a highly accurate value can be difficult. In Figure 19.9 we see a table showing that even after 400 terms we can only comfortably predict that the value is perhaps between 0.78 and 0.79. It turns out that that the true value is $\pi/4 = 0.785398\ldots$ This amazing fact can be shown by finding the Taylor series for arctan(x) and recalling that arctan(1) = $\pi/4$.

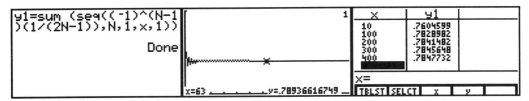

Figure 19.9 A slow converging graph and table.

20. GEOMETRIC SERIES

In the last chapter, we found Taylor polynomials by starting with a function and forming the successive sums to find a series representation. In some cases, there is the reverse situation: the series is known, but the function is not. We now consider finding the function from a given series. The graphing calculator will, of course, be of little help in writing a general formula for a sum, but it can help clarify the generalized sum and verify with finite sums.

The general formula for a finite geometric series

A finite geometric series has the form

$$a + ax + ax^2 + \ldots + ax^{n-1} + ax^n$$

Note that the coefficient a is the same for each term. The closed form sum is

$$S_n = \sum_{i=1}^{n} ax^{n-1} = a\frac{(1-x^n)}{(1-x)}, \quad (x \neq 1)$$

You might think that such a sum would be so rare as to make it not worth considering, but this kind of finite series comes up in many situations.

Repeated drug dosage using sum(seq(...))

Consider a 250 mg dose of an antibiotic taken every six hours for many days. The body retains only 4% of the drug in the body after six hours. The interesting part is that this does not say that 4% of 250 mg is left. This is only true at the end of the first six hours. At the end of the second six hours, the body retains 4% of the second dose *and* 4% of the remaining first dose. Let's make a sequence of the amount of drug in the body right after taking the nth dose.

$$Q_1 = 250,$$

$$Q_2 = Q_1(0.04) + 250,$$

$$Q_3 = Q_2(0.04) + 250, \text{ etc.}$$

But if we substitute lower sums and multiply, we find

$$Q_2 = 250(0.04) + 250 \text{ and}$$

$$Q_3 = 250(0.04)^2 + 250(0.04) + 250$$

In general,

$$Q_n = 250(0.04)^{n-1} + \ldots + 250(0.04)^2 + 250(0.04) + 250.$$

20. GEOMETRIC SERIES   117

The Q's are a finite geometric series with $a = 250$ and $x = 0.04$. Let's calculate the amount of antibiotic in the body at the time of taking the fourth pill. In Figure 20.1 we use `seq` to create a list showing the amount from each pill that is left. The total is found by summing the sequence. Finally, we check our answer by using the formula and the results are the same.

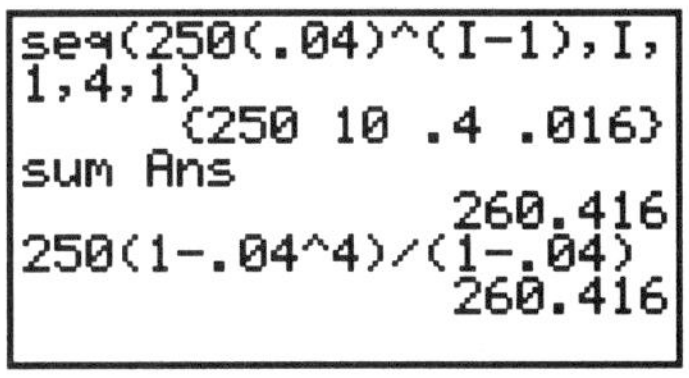

Figure 20.1 The dosage present in the body at the fourth pill summed as a sequence and compared to the formula.

Using a table to find consecutive finite sums of a sequence

To see consecutive finite sums of a sequence, you can define a function using `sum(seq(...))` and set up a table to show the desired values. This is shown in Figure 20.2

Figure 20.2 Showing finite sums using a function.

You might observe that at the end of the first day (the fifth pill), your medication is close to a stable amount. A different drug might have a much higher retention level. Changing 0.04 to 0.50 gives the table in Figure 20.3 which shows that a drug with this level will take more than two days to reach its stable limit of about 500 mg.

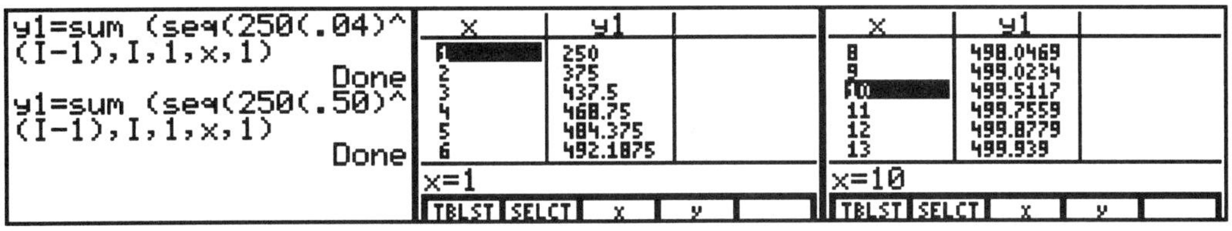

Figure 20.3 Retention amounts for a rate of 50%.

Regular deposits to a savings account

Another direct application of the finite geometric series is the value of an investment that earns interest. Suppose that you plan for retirement by putting $1000 a year into a savings account that earns 5% annual interest. You might want to know how much this will be worth after the nth deposit. We will find this using two methods in Figure 20.4. First, define a y function and make a table showing the value every ten years after the first year. Second, in Figure 20.4, we will just use the finite sum formula to check the sum when $n = 41$, when you might start thinking about retirement.

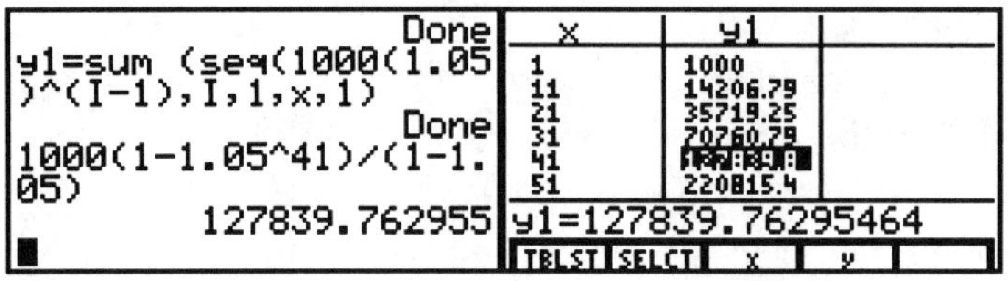

Figure 20.4 The value of a 5% savings account after 41 annual deposits.

Identifying the parameters of a geometric series

When we see several sums like $a + ax + ax^2 + \ldots + ax^{n-1} + ax^n$, it is helpful to have a generalized definition of y1. In Figure 20.5 we do this with a changed to the capital A and x renamed B. Beware that in this definition x is the n — there is no way to avoid this confusing assignment of variables since a y definition must have X as the independent variable. We check our new definition with the first antibiotic retention model.

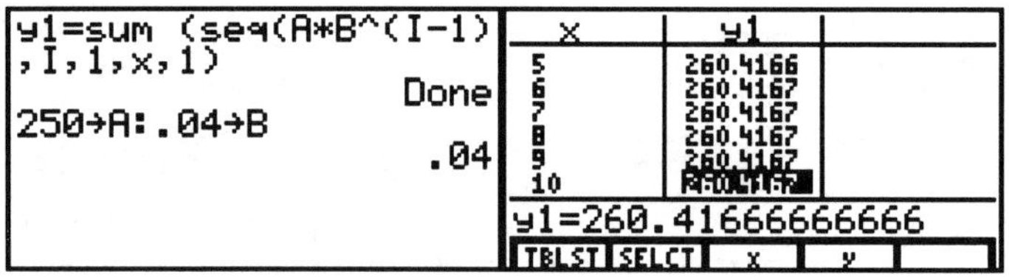

Figure 20.5 Checking to see that a general method gives the same results as before.

▶ *Tip: Remember that upper case letters are easier to use as variable names since the* STO→ *key puts you in the* ALPHA*-lock mode.*

Now we are ready to practice on the following list of infinite series.

(a) $1 + \dfrac{1}{2} + \dfrac{1}{4} + \dfrac{1}{8} + \cdots$

(b) $1 + 2 + 4 + 8 + \cdots$

(c) $6 - 2 + \dfrac{2}{3} - \dfrac{2}{9} + \dfrac{2}{27} - \cdots$

The hardest part is identifying the values of A and B in each case. By entering these and checking a table, you can quickly find that series (a) converges to 2, (b) diverges, and (c) converges to 4.5.

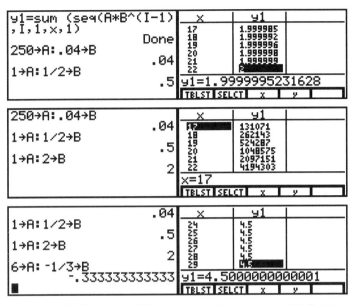

Figure 20.6 Checking different series from a general definition.

Summing an infinite series by the formula

Returning to the antibiotic example, we could (hypothetically) take a dose every six hours forever and then the series would be infinite. We have seen infinite sums that add up to finite values, such as the 'half a cookie diet' example of Chapter 17. The infinite sum of a geometric series is given by the following formula:

$$a + ax + ax^2 + \ldots + ax^{n-1} + ax^n + \ldots = \frac{a}{1-x}, \quad \text{for } |x| < 1$$

In Figure 20.7 we revisit the formula for the two drug dosage cases we investigated as finite sums in Figure 20.2 and 20.3. They both qualify as infinite sums since each x-value ($x = 0.04$ and $x = 0.5$) satisfies the condition $|x| < 1$. When we use the formula to sum the series, we see that it is indeed the table limits. We can't expect to sum all infinite series; in the previous example, series (b) had $x = 2$ and did not converge.

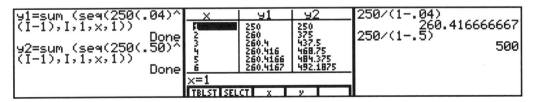

Figure 20.7 Finite sums using a table and verifying the infinite sums with the formula.

Piggy-bank vs. trust

Suppose that your parents are trying to decide on a plan to provide for your future and they have two choices:

  I. Each year they put your age in dollars into a piggy bank.

  II. Each year they put $3 into a trust account that earns 6% annual interest.

We already know how plan II works: use $A = 3$ and $B = 1.06$ in the formula above. In plan I we need to look at the sum of the series $1 + 2 + 3 + 4 + \ldots + n$. This is *not* a geometric series. A clever way to consider this sum is to write it twice, once forward and once backward:

$$
\begin{array}{r}
1 + 2 + 3 + \ldots + n-1 + n = S_n \\
n + n-1 + n-2 + \ldots + 2 + 1 = S_n \\
\hline
n+1 + n+1 + n+1 + \ldots + n+1 + n+1 = 2S_n
\end{array}
$$

There are n of these $(n+1)$ sums, so we have $S_n = n(n+1)/2$. We enter this formula in y2 and look at some early values in a table in Figure 20.8. It seems that the piggy bank is the better plan. But as you get to 65 years old, we see that the two plans seem about even, and as you get into your seventies, plan II is much better. This comparison is graphically clear from the graph in the bottom row of Figure 20.8.

▶ *Tip: (TI-86 users) Graphing a* `sum(seq(...))` *function is extremely slow as x increases — you may want to set* xRes=10.

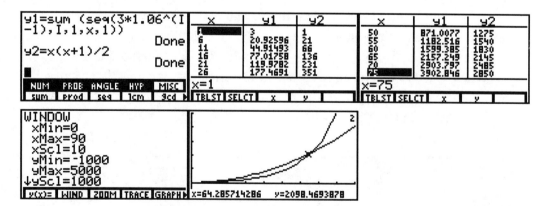

Figure 20.8 The trust is better than the piggy bank after age 65.

21. FOURIER SERIES

The Taylor polynomials are good approximations to a function, but beyond the radius of convergence they're awful. As suggested by the example of the sine function in Chapter 19, Taylor polynomials are practically useless for periodic functions. We now find a set of approximating functions that globally approximate a periodic function. These global approximation functions are called Fourier approximations. First we will discuss how to graph some periodic functions that are not trigonometric, including piecewise defined functions.

Periodic function graphs

The only kinds of periodic functions we have graphed so far are trigonometric. We now introduce and graph a few other types.

The rise and crash function: y1 = x-int(x)

Our first example is $y = x - \text{int}(x)$, a linearly increasing function that falls back to 0 when it reaches a height of 1. It is defined using $f(x) = \text{int}(x)$, the greatest integer

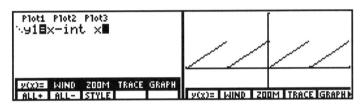

Figure 21.1 A non-trigonometric periodic function in the window $-2 \leq x \leq 2$, $-1 \leq y \leq 2$.

function. Recall that it is in the catalog or it can be found in the 2nd_MATH NUM menu (or you can just spell it out, remembering to use lower case letters).

By using the dot graph style to graph this type of function, you can avoid the jagged connected graph problem.

▶ *Tip: Use the GRAPH MORE FORMT command and change the global graph mode to DrawDot. This prevents the unwanted connecting of values over discontinuities, which arise frequently in piecewise defined functions.*

The square wave function: using a logical expression (x<0)

This function is commonly used in electrical engineering to model switching: it is either on or off. We use 1 to mean on, 0 to mean off. In this case we will graph one period of the function using the logic feature of the TI.

If we evaluate the logical expression (0<x), then it will be TRUE or FALSE depending upon the current value of x. The arithmetic value of TRUE is 1, of FALSE is 0. Thus an expression like (0<x) has a value of one if $x < 0$ and zero otherwise. Inequality symbols, such as < and ≤, are pasted in from the

2nd_TEST menu. We can actually make a logical expression into a function and graph it; see Figure 21.2.

If we multiply by another logic expression, say (x≤1), then the product (0<x)(x≤1) will have a value of one only for $0 < x \leq 1$, and be zero otherwise. The graph of this expression is shown in Figure 21.2

This second definition reproduces the square wave function of period 2 for a complete period, $0 \leq x \leq 2$. Later in the chapter, we will approximate and graph Fourier approximations to this function.

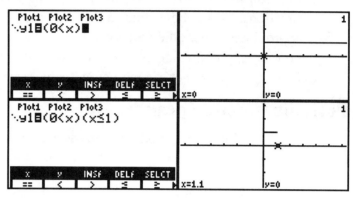

Figure 21.2 *A logical expression in a function definition.*

Piecewise defined functions

This technique of putting logical expression in function definitions can be used as a means to enter a piecewise defined function: multiply the appropriate function rule by the logical expression for the interval on which it is defined. For example, look at the function definition of y1 in Figure 21.3. It is the squaring function on the interval $0 < x \leq 1$ and zero otherwise. To create more complicated piecewise defined functions, add an expression for each interval of definition. We show an example of this in the next section.

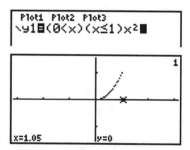

Figure 21.3 *A logical expression restricting the function definition (using ZDECM settings divided by 2).*

The triangle wave function

A complete period of the triangle wave function starts at 0, increases linearly to one at $x = 1/2$, then decreases linearly to zero again at $x = 1$. The two pieces of this function can be added together to make a single function definition as shown in Figure 21.4. We could also make a more complicated definition that would actually repeat periodically, but we only need one period for the Fourier approximation function.

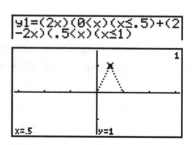

Figure 21.4 *A single period of the triangle wave function.*

The general formula for the Fourier approximation function

A Fourier approximation function uses the sine and cosine functions to approximate periodic functions. These functions are not polynomials, but we use the polynomial vocabulary to describe them. We use the term *degree* to specify which sine/cosine terms are included and specific constant multipliers are called the *coefficients*. Here is the definition of the nth degree Fourier function for the interval $-\pi \leq x \leq \pi$:

$$F_n(x) = a_0 + a_1 \cos(x) + a_2 \cos(2x) + \ldots + a_n \cos(nx)$$
$$+ b_1 \sin(x) + b_2 \sin(2x) + \ldots + b_n \sin(nx)$$

With this basic structure in mind, we give the generalized definition which includes the general period b and the definitions of the coefficients.

$$F_n(x) = a_0 + a_1 \cos((\tfrac{2\pi}{b})x) + a_2 \cos(2(\tfrac{2\pi}{b})x) + \ldots + a_n \cos(n(\tfrac{2\pi}{b})x)$$
$$+ b_1 \sin(x(\tfrac{2\pi}{b})) + b_2 \sin(2(\tfrac{2\pi}{b})x) + \ldots + b_n \sin(n(\tfrac{2\pi}{b})x)$$

$$a_0 = \frac{1}{b}\int_{-b/2}^{b/2} f(x)dx$$

$$a_k = \frac{2}{b}\int_{-b/2}^{b/2} f(x)\cos(k(\tfrac{2\pi}{b})x)dx \quad \text{for } k > 0$$

$$b_k = \frac{2}{b}\int_{-b/2}^{b/2} f(x)\sin(k(\tfrac{2\pi}{b})x)dx \quad \text{for } k > 0$$

A program for the Fourier approximation function

If there was ever a task calling for programming, this is it. We use a program called FOURIER to do the work for us. If possible, find a copy that you can load through the LINK.

Fourier program listing

```
ClLCD
Disp "Put function in y1."
Disp "Set nice window."
Disp "------------------"
Disp "It lists coefficients"
Disp "in LC and LS."
Disp "FPolys in y2-y9 "
Input "Go On? (0=N,1=Y) ",N
If N<1
Stop

Func
9→dimL LC
9→dimL LS
1E-5→tol
```

```
ClLCD
Input "Period length? ",B
2π/B→C
Disp "----------------"
Disp "Takes 2 minutes."
For(K,1,8,1)
(2/B)fnInt(y1*cos (K*C*x),x,0,B)→LC(K)
(2/B)fnInt(y1*sin (K*C*x),x,0,B)→LS(K)
End
y2=(1/B)fnInt(y1,x,0,B)+LC(1)*cos (C*x)+LS(1)*sin (C*x)
y3=y2+LC(2)*cos (2C*x)+LS(2)*sin (2C*x)
y4=y3+LC(3)*cos (3C*x)+LS(3)*sin (3C*x)
y5=y4+LC(4)*cos (4C*x)+LS(4)*sin (4C*x)
y6=y5+LC(5)*cos (5C*x)+LS(5)*sin (5C*x)
y7=y6+LC(6)*cos (6C*x)+LS(6)*sin (6C*x)
y8=y7+LC(7)*cos (7C*x)+LS(7)*sin (7C*x)
y9=y8+LC(8)*cos (8C*x)+LS(8)*sin (8C*x)

Lbl N1
ClLCD
Disp "---- (0 = Quit) ----"
Input "Graph Poly? (1-8) ",N
If N<1
Stop

FnOff :FnOn 1,N+1
DispG
Pause
Goto N1
```

Setting up before using FOURIER

Enter the function in y1 with a definition that will span one complete period starting at zero. (This is why we only needed one period of the square wave and triangle wave functions.) Find a nice window for y1 that would show several periods on each side of the origin. Check the lists LC and LS ('list cosine' and 'list sine') to be sure they don't contain data that you want — they will be erased. All the function definitions from y2 to y10 will be erased as well and used to store the 'polynomials' of the various orders.

Following the prompts after pressing PRGM FOURIER

We start by using the FOURIER program to create and graph the Fourier approximation function of the square wave function. This sequence is shown in Figure 21.5, and the graphs are shown in Figure 21.6. The GRAPH MORE FORMT setting has been changed back to DrawLine to better show the Fourier functions. First define the function y1 and set a nice window that will show several periods (we use ZDECM divided by 2). On a fresh line of the home screen, press PRGM NAMES (F1) and select FOURIER. Start the program by pressing ENTER. The opening screen is just a reminder that y1 should be the original function. If you have done a setup, then press 1 ENTER to continue.

Next you will be prompted to enter the period. It will then take about two minutes to calculate and store the coefficients in LC and LS. The final prompt is a choice on which order of polynomial you would like to graph. When the graph is drawn — this can also take a minute — it will pause. Press ENTER and it will allow you to quit (0=Quit) or graph another Fourier function.

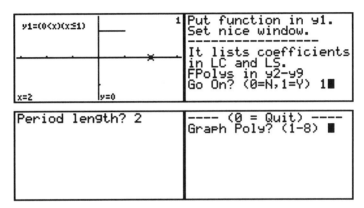

Figure 21.5 Setting up and using the FOURIER program on the square wave function.

The graphs produced by PRGM FOURIER

In Figure 21.6 you can see three different graphs approximating the square wave function. They are of orders 3, 5 and 7; as you see the fit gets better as the order increases. All values in LC are essentially zero, meaning it does not use the cosine function to approximate this function. Only the LS odd entries contribute non-zero values, so increasing the order from 3 to 4 will result in the same Fourier polynomial.

➤ *Tip: Remember that it is only required to define and graph one complete period of the original, yet the polynomials are defined on all real numbers.*

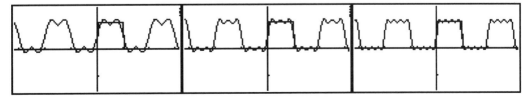

Figure 21.6 Square function Fourier fits of degrees 3, 5 and 7.

Fitting the triangle wave function

We close with an approximation of the triangle wave function as defined in Figure 21.4. It has a period of 1 and the polynomials are a close fit, even with degree 2. The sine curve shape is closer to that of the triangle wave than the square wave, so it makes sense that the fit is very good even at degree 2.

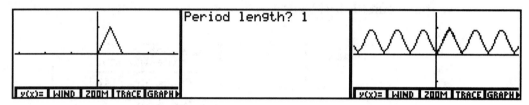

Figure 21.7 A Fourier function of degree 2 fits the triangle wave function very closely.

Periodic function definition (optional)

Because only one period is needed to define a Fourier polynomial, the periodic functions stored in y(x)= have, for simplicity, been defined only for one period. To define a function that graphs the complete square function of period 2, use

$$y1 = (\text{mod}(x,2) \leq 1)$$

The mod (modulo) function divides the first expression by the second entry and returns the non-negative remainder. Thus mod(x,p) will produce values that repeat with period *p*. We can use this idea to enter the complete triangle function. Replace the references to x with mod(x,1) and break the definition into two logical parts:

$$y1 = 2\text{mod}(x,1)(\text{mod}(x,1) < .5) + (2 - 2\text{mod}(x,1))(\text{mod}(x,1) \geq .5)$$

➤ *Tip: The FOURIER program is already slow and is intolerably slow when using this full version of the triangle wave function. Making the y1 function as simple as possible, like using a one period definition, helps the computation time.*

PART V
DIFFERENTIAL EQUATIONS

22. DIFFERENTIAL EQUATIONS AND SLOPE FIELDS
23. EULER'S METHOD
24. SECOND-ORDER DIFFERENTIAL EQUATIONS
25. THE LOGISTIC POPULATION MODEL
26. SYSTEMS OF EQUATIONS AND THE PHASE PLANE

22. DIFFERENTIAL EQUATIONS AND SLOPE FIELDS

Often a situation occurs where we know something about a rate, but the original function may not be explicitly known. For example, we have seen how to antidifferentiate the equation $\frac{dQ}{dt} = t$ and solve for Q to find $Q = t^2 + C$. This is a simple case. It could be that the rate might depend on Q; the antidifferentiation is not so straightforward then. Equations such as these that involve derivatives are called differential equations.

As in the above example, t is most often the independent variable, as most rates are expressed in terms of time. The dependent variable is often a quantity, so t and Q are the variables used to enter differential equations on the TI-86 and TI-85. It is also quite common in particularly in textbooks to use the standard x and y variables, with a prime denoting a derivative. For example, the above differential equation could also be written more succinctly as $y' = x$.

A word about solving differential equations

In some cases, there are analytic solutions to differential equations, such as the example above. The current TI Eighty series calculators only solve equations numerically. This means that they cannot tell you the solution as a formula, but they can give you a table or a graph of the solution. This means that the calculator is most useful in checking to see if your analytic solution agrees with the numerically derived solution graph or table. The numerically derived solution can also be helpful in situations where you are doing guess-and-check to find an analytic solution. Looking at a graph will help you guess the type of solution function.

General vs. particular solutions

In the above example, the analytic solution was written with the constant C. This means that there is a whole set of solutions which differ by a constant. To graph this whole set would fill the screen. A graph to show local representative behavior, called slope lines, is called a slope field graph. The main purpose of this chapter is to introduce this tool.

However, we often know specific conditions that determine a single solution, called a particular solution. Particular solutions are graphed as curves. It is common to draw a slope field graph to show the general solution and then superimpose a particular solution curve starting at some initial value.

Discrete vs. continuous representation

Some calculators, such as the TI-85 and TI-86, are blessed with a differential equation mode, where the continuous differential equation can be directly entered and graphed. Other calculators, such as the TI-83, do not have such a mode, but they can enter and graph difference equations — a kind of discrete differential equation. The two approaches, the discrete and continuous, offer insight into the topic when taken together. We now see an example of how a discrete difference equation approximates a differential equation.

The discrete learning curve using lists

One theory of learning is that the more you know, the slower you will learn. Let y be the percentage of a task we know. The learning rate is then y' and we assume $y' = 100 - y$. The time unit for this rate is the standard five day work week. The slowing of learning takes place immediately and continuously, but we will start by looking at a discrete model where we assume the same rate all day. This assumption makes the model discrete. In *Calculus*, by Hughes-Hallett, et al., we find the following table:

Time (working days)	0	1	2	3	4	5	10	20
Percentage learned	0	20	36	48.8	59.0	67.2	89.3	98.8

Figure 22.1 Approximate percentage of task learned as a function of time.

Consider a new employee who knows 0% of a task at time 0, Monday morning. She learns at the rate $y' = (100 - 0)\%$ during the first day. The part of the task the employee learns on Monday, one fifth of a work week, is

$$y' * (1/5) = 100\%(0.2) = 20\%$$

At this rate, she would have the task entirely mastered in a week. But Tuesday, because she already knows 20% of the task, her learning will slow to

$$y' = 100\% - 20\% = 80\%$$

Therefore, on Tuesday the part of the task she learns is an additional

$$y' * (1/5) = 80\%(0.2) = 16\%$$

In total, she has learned 20% + 16% = 36% by the end of Tuesday.

Making a list with a loop: `For()...End`

To create our lists we will use the `For()` command. The general structure is:

For(index, start, stop, increment)

which is similar to the `seq()` function. The increment is optional; the default setting is 1. Each `For()` must be paired with `End`. The set of commands between `For()` and `End` will be repeatedly done until the index exceeds the stop value.

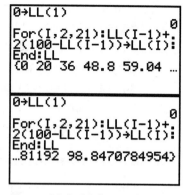

Figure 22.2 Making a list to show values of the discrete learning function and checking the last value by scrolling.

We show in Figure 22.2 how to create the values shown in the table of Figure 22.1. The basic method is to

- set the initial value in the list,
- generate successive values using a `For()` ... `End`,
- display the list of mastery values of the learning function.

Notice that the index starts at 2, since our initial value is in `LL(1)`. It would have been clearer in this example to start with `LL(0)`, but the TI does not allow zero as a list index. We will get around this with the following column view of lists.

▶ *Tip: A long entry like the one in Figure 22.2 can be entered and saved as lines in a program, then altered to meet the current situation before executing the program.*

The horizontal listing is awkward, so we use **2nd_STAT EDIT** to display the values in columns. (This is only available on the TI-86.) By moving to a vacant heading, the list names can be entered and the data will appear in the columns underneath. By deleting the top zero in `LL`, the index matches the table in Figure 22.1. We can interpret the list values as saying that the employee learns only 67% (`LL(5)`) of the task by the end of the week. By the end of her second week, the new employee knows 89% (`LL(10)`) of the task, and by the end of four weeks, 99% (`LL(20)`). According to the model, it is impossible to completely learn the task.

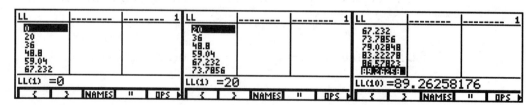

Figure 22.3 Viewing lists from **2nd_STAT EDIT**. *The first zero is deleted to match the index from Figure 22.1.* `LL(10)` *is the amount learned at the end of ten days.*

The continuous learning curve $y' = 100 - y$

The discrete version of the learning curve assumed that the worker had the same learning rate all day. Now suppose that the learning rate changed every instant, i.e., that it is a continuously changing rate. The slope field and a particular solution of this differential equation are shown in Figure 22.4.

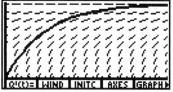

Figure 22.4 *The learning curve slope field and a particular solution with y(0)=0.*

▶ *Tip for TI-85 users: Your calculator has no built-in slope field command and the related commands are somewhat different for the TI-85. We suggest that you skim over the keystroke sequence, but look carefully at the graphs. At the end of this chapter is a TI-85 section which includes a program to draw slope fields.*

How did we get this impressive graph? There are numerous leading to it because there are so many options and settings. Many of the settings will be at default and need not be changed. For differential equation graphing, it is a good idea to use a prescribed order to set up and graph. This order is easy to learn: first set the differential equation mode and its format, then follow the order of the five GRAPH menu keys.

Use 2nd_MODE to change the graphing mode on the fifth row to the differential equations setting. (You must press ENTER to select DifEq before you exit.)

Press GRAPH MORE FORMT to be sure all settings are at the default (left-most). It is important that the bottom row be set before entering equations. We now follow the order of the F1 to F5 keys.

Press F1 (Q'(t)=) and enter the equation definition. Be careful to enter *y* as Q1, not just Q. Use t as *x*, if it is involved. (The default graph style is bold for differential equations.)

Press EXIT F2 to set the window. The settings are shown in two partial screens. The *x*-values are in weeks. The *y*-values will show up to 100%.

132 PART V / DIFFERENTIAL EQUATIONS

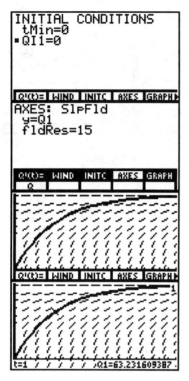

Use F3 (INITC) to set QI1= to the initial conditions of Q1 for the particular solution. If you use CLEAR and leave QI1 blank, then no particular solution will be drawn, just the slope field.

Use F4 (AXES) to be sure the *y*-axis is Q1. For the SlpFld setting, the *x*-axis is automatically *t*. The fldRes setting is the number of columns of slope lines drawn in the slope field.

Press 2nd_F5 (GRAPH). The slope field shows us that the values appear to be converging to 100 as *t* gets larger — this is independent of where *t* starts.

Press MORE TRACE to provide numerical data for the particular solution. By tracing to t=1 (1 week) we see that Q1(1) is about 63% This is slightly lower than the LL(5) ≈ 67%, the discrete value at the end of the first week (5 days.) For Q1 the learning was continuous and slowing all day long.

Slope fields for several differential equations

We now show several examples of setting up and graphing slope lines for various differential equations. We assume that the MODE, FORMT and AXES settings are as in the previous example.

Slope field for *y'* = *y*

The solution to this differential equation is the exponential function $y = e^x + C$. In Figure 22.5 we start by entering the differential equation as Q'1=Q1. Next we set an initial condition, QI1=1. We bypassed WIND because we set a window this time by pressing MORE ZOOM MORE ZDECM. The particular solution, $y = e^x$, is shown starting at (0,1).

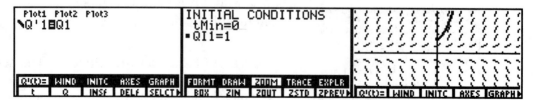

Figure 22.5 Slope curves for the solution $y = e^x + C$, using ZOOM DECM.

Slope field for $y' = 2x$

The general solution of $\frac{dy}{dx} = 2x$ is $y = x^2 + C$. As we can envision from the slope field, a particular solution will be a parabola whose vertex is on the y-axis. By clearing the initial conditions, i.e., leaving QI1 blank, no particular solution is graphed. The first graph in Figure 22.6 is the slope field of this equation. We can still add graphs of particular solutions by pressing MORE EXPLR (F5). This activates a small blinking cross hair that can be moved to the point on the screen where you want a particular solution to begin. In the second graph, we moved the cursor to (-1,1) and pressed ENTER which drew (part of) the particular solution $y = x^2$. By pressing EXIT, the EXPLR menu item is available again and can be used to add other explorations to this same graph.

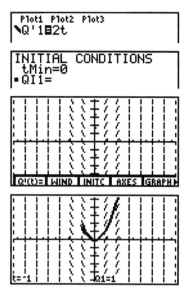

Figure 22.6 Slope curves showing no particular solution. Then EXPLR is used.

Slope field for $y' = -x/y$ and $y' = x/y$

If you follow a set of slope lines in the first graph of Figure 22.7, you will trace out a circle centered at the origin. By implicit differentiation, we know that this differential equation is obtained from the equation of a circle, $x^2 + y^2 = a$ where $a > 0$.

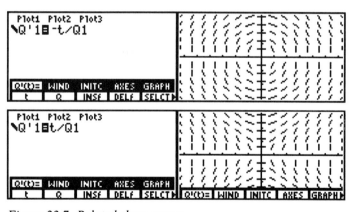

Figure 22.7 Related slope curves.

A sign change from the previous differential equation produces a new slope field in which we could trace out hyperbolas. Again by implicit differentiation, we know that this differential equation is obtained from the equation of a hyperbola of the form $x^2 - y^2 = a$ where $a > 0$. With a modest leap from this and the previous subsection, you can see that all conic sections can be defined in terms of differential equations.

Slope field for a predator-prey model $y' = (-y + xy) / (x - xy)$

The final example is a slope field of a differential equation that relates two interacting populations (in particular, one population eats the other). Since we only care about positive populations, we set xMin=0 and yMin=0 in the ZDECM screen. A particular solution is an oval curve, as the populations are cyclical, unless we are at (1,1), a fixed point. This kind of equation will be discussed in Chapter 26.

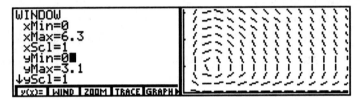

Figure 22.8 Slope curves of the predator-prey model `Q'1= (-Q1+t*Q1)/(t-t*Q1)`

➤ *Tip: This concludes the chapter material for the TI-86 users. TI-85 users who want the capability to graph slope fields should continue.*

A program for slope fields on the TI-85

The following program can be entered or copied from another TI-85 (using LINK) that already has it entered. This was slightly adapted from the one published in the TI magazine *Eightysomething!,* Spring 1995. It can be seen at the internet address

http://www.ti.com/calc/docs/act/pdf/eighty3.pdf

It will draw the slope field for a differential equation of the form $y' = F(x, y)$. The indentation in the listing is to help make the program more readable; each line must actually be entered at the left margin, without leading spaces. Name the program SLPFLD.

Slope field program listing

```
ClLCD
FnOff
8→A
12→B
(yMax-yMin)/A→V
(xMax-xMin)/B→H
For(I,1,A)
 For(J,1,B)
  yMin+V(I-1)+V/2→y
```

22. DIFFERENTIAL EQUATIONS AND SLOPE FIELDS   135

```
  xMin+H(J-1)+H/2→x
  y1→M
  y-M*.3H→S
  y+M*.3H→Z
  x-.3H→P
  x+.3H→Q
  If abs(Z-S)>.6V:Then
    y+.3V→Z
    y-.3V→S
    .3(Z-y)/M+x→Q
    .3(S-y)/M+x→P
  End
  Line(P,S,Q,Z)
 End
End
```

Setting up before using SLPFLD with a particular solution

The particular solution should be graphed first using the following sequence that is similar to that used for the TI-86 and shown in Figure 22.9:

- 2nd_MODE: Change MODE to DifEq
- F1: Enter equations in Q'(t)= (use Q and t as variables)
- F2: Set the window
- F3: Set the initial conditions
- F4: Check axes
- F5: Graph the particular solution
- On the home screen, define the equation as y1 (use x and y as variables)
- Execute SLPFLD to add the slope field to the graph of the particular solution

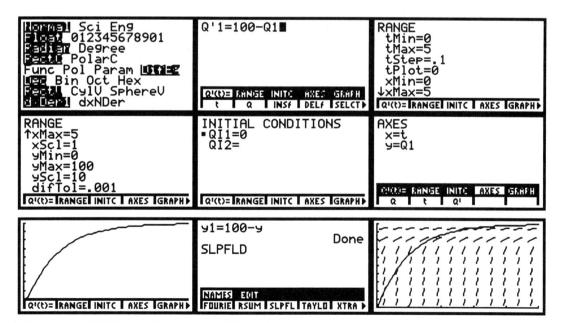

Figure 22.9 TI-85 steps for a slope field graph with a particular solution.

Using SLPFLD without a particular solution

Before using the SLPFLD program to show slope lines of a differential equation of the form $y' = F(x, y)$, you must place the $F(x, y)$ side of the equation in y1. This means that y1 will be an expression using the variables x and y. Notice that the Q and t notation from the Q'(t)= definition must be translated to y and x when entered on the y(x)= screen. In the previous example, you made this definition on the home screen and then ran the program SLPFLD.

If no particular solution is to be graphed, you do not need to be in DifEq mode. The function can be entered either from the y(x)= screen or from the home screen (using x and y).

Next, set what you hope is an appropriate window. It is not always possible to know an appropriate window ahead of time, so you may need to use a trial-and-error approach. Note, however, that a y1 function that includes the variable y may not graph properly if you press GRAPH. It is still the case that y(x)= functions can only have x as the variable for normal graphing; the SLPFLD program gets around this by not using the GRAPH mode. Thus, you must use the SLPFLD program to find a nice window; pressing the GRAPH key may result in an error message screen. A typical setup and graph are shown in Figure 22.10.

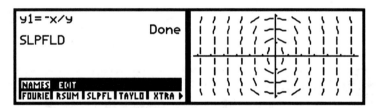

Figure 22.10 TI-85 slope field graph in a ZDECM window without a particular solution.

23. EULER'S METHOD

In the last chapter we saw how the graph of a particular solution follows the slope lines. The method of graphing by starting at an initial point and following short linear slope lines is called Euler's method. This technique was used in the last chapter to find the discrete values that approximated the particular solution of the learning curve differential equation. To increase accuracy, we can shorten the slope lines.

The relationship of a differential equation to a difference equation

Suppose we want to find the solution of the differential equation

$$\frac{dy}{dx} = F(x, y) \text{ starting at the point } (x_0, y_0)$$

We think about this continuous differential equation in a discrete way and write

$$\frac{\Delta y}{\Delta x} = F(x, y)$$

or, more explicitly,

$$\frac{y_n - y_{n-1}}{\Delta x} = F(x_{n-1}, y_{n-1}) \text{ with } \Delta x = x_n - x_{n-1}$$

Now multiply both sides by Δx and rearrange in a form we can use,

$$y_n = y_{n-1} + F(x_{n-1}, y_{n-1})\Delta x \qquad x_n = x_{n-1} + \Delta x \qquad (*)$$

From this, you can see how knowing (x_0, y_0) allows us to find y_1. We will then use y_1 to find the value of y_2, etc.

Translating a differential equation to a difference equation for y' = y starting at (0,1)

Consider the differential equation $\frac{dy}{dx} = y$ starting at (0,1). With $\Delta x = 0.1$, we make a table using Euler's method and then make a graph to compare it to our analytic solution. We use LY to store the y-values. The y_n formula marked (*) above translates to

```
LY(N)=LY(N-1)+LY(N-1)(0.1)=1.1*LY(N-1)
```

We want to compare values on the interval $0 \leq x \leq 1$, so with $\Delta x = 0.1$ we will need to calculate 10 values. We accomplish this by using a `For():End` pair; see the first screen of Figure 23.1.

Making a step function from the list

The list LY can be displayed using the STAT EDIT feature, but instead we define a step function from LY. (Before starting, be sure the graphing mode is set to the default, Func.) To define the list LY as a step function in y1, we must insure that the index is an integer. To do this we multiply *x* by 10, take the integer part, and add 1 (this gives 1,2,3,...). This cheap trick is simpler that using a data plot, which we will do in Chapter 25. We set y2 to be our analytic particular solution, $y = e^x$. We can compare the numeric approximations by using TABLE. However, you must be careful not to scroll beyond $0 \leq x \leq 1$, else an error will occur. We observe the same precaution in the window setting as we graph the two functions. The point here is that the Euler method approximations sketch a curve that approximates the solution. You can see that, at a calculated *x*-value such as .5, the approximation is good. If needed, these approximations can be improved by making Δx smaller.

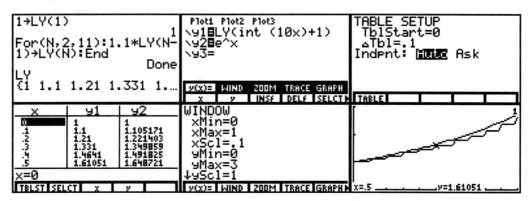

Figure 23.1 Making a list for Euler's method of approximating a particular solution of the differential equation $y' = y$. A table (TI-86 only) and graph compare the approximation.

Euler's method for *y'* = -*x* / *y* starting at (0,1)

Consider the differential equation $\frac{dy}{dx} = \frac{-x}{y}$ starting at (0,1). We want to investigate the effect of using a smaller value of Δx by graphing the solution on the interval $0 \leq x \leq 1$ first with $\Delta x = 0.1$, then with $\Delta x = 0.01$. We could repeat the previous method (which TI-85 users will have to do), but the TI-86 has a built-in Euler setting. The setup process follows the steps outlined in the last chapter.

23. EULER'S METHOD   **139**

In MODE, change to DifEq. Then press GRAPH MORE FORMT and change the bottom two settings to Euler and FldOff. (The TI-85 does not have these settings, but you can obtain similar graphs by following this sequence.)

Press GRAPH F1 and enter the equation.

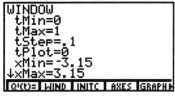

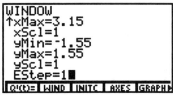

Press EXIT F2 to set a window. The x- and y-values are from ZDECM divided by two — see the tip below for an easy way to do this. We are interested in $0 \leq t \leq 1$, but we make x larger for a wider window.

The number of iterations between each tStep is given by EStep; one is its default. In the next example, we increase EStep to 10 to improve the accuracy of the approximation.

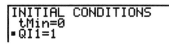

Press F3 : (INITC) and set QI1=1.

Press F4 : (AXES) to check axes.

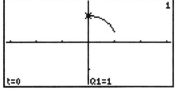

Use TRACE to find the points (0, 1), (0.1, 1), (0.2, 0.99), (0.3, 0.97), etc.

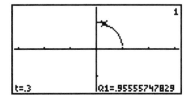

Changing EStep to 10 gives much better results: (0, 1), (0.1, 0.995), (0.2, 0.981), (0.3, 0.956), etc. The graph is more circular as it should be (recall the analysis of this differential equation in the previous chapter).

➤ *Tip: A quick way to get a nice window is to begin with ZDECM and improve it using ZIN or ZOUT. Under ZFACT, you can set xFact=2 and yFact=2 for halving and doubling window settings.*

Euler gets lost going around a corner

Use Euler's method with caution. As you wander away from your initial point, you may encounter increasing errors. The next example shows that there are some paths where we encounter an infinite slope that will cause problems.

Let's take the previous example and set **tMax** to four in hopes that it will sweep through the *x*-axis and continue to follow its circular path for a complete circle. Figure 23.2 shows the folly as we come to (1,0). Changing the setting from **Euler** back to the default **RK** eliminates this hazard (available only on the TI-86). The setting **RK** stands for Runge-Kutta method, which is slower than the Euler method but more accurate (see any differential equations text for details).

➤ *Tip: Should you desire a graph of the equation in the fourth quadrant, you can set the initial value as* **QI1=-1**.

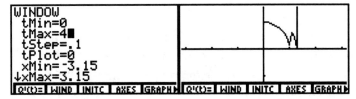

Figure 23.2 Euler's method breaks down as at (1,0) where the slope is undefined.

24. SECOND-ORDER DIFFERENTIAL EQUATIONS

A second-order differential equation has a second derivative involved in the equation expression. Before making the transition from a symbolic equation to necessary calculator definitions, we rewrite these equations by solving for the second derivative. Thus we will find it easiest to consider a differential equation in the form $y'' = F(x, y, y')$.

As mentioned before, this TI calculator will not give symbolic solutions of differential equations, but it can be used to get numerical solution values and to check your analytical solution by graphic means. We begin with the simplest second order equation.

The second-order equation $s'' = -g$

Figure 24.1 The TI does not do nested integration: error message from trying to graph in **Func** mode.

This is a classic equation from physics, $\frac{d^2s}{dt^2} = -g$, where g is the constant force of gravity on a falling object, s is displacement in feet, and t is time in seconds. We assume g is 32 ft./sec^2, the initial velocity is zero ($v_0 = 0$), and the initial distance is zero ($s_0 = 0$). Antidifferentiating once gives $s' = -32t$, and then again gives the solution $s = -16t^2$.

If we did not know these two simple antiderivatives, we might have tried to use the calculator to apply numerical integration twice,

$$y1=\text{fnInt}(-32,x,0,x)$$
$$y2=\text{fnInt}(y1,x,0,x)$$

and graph **y2** as our solution. This cannot be done on this calculator, as shown in Figure 24.1.

There is also no **Q''(t)=** menu item or other feature for entering a second-order equation, but we can build a second order equation as a system of two first order equations:

$$\mathtt{Q'1=Q2} \text{ and } \mathtt{Q'2} = \text{-32 gives}$$

$$(Q1)'' = (Q'1)' = (Q2)' = Q'2 = \text{-32}$$

For our first example, we give a detailed sequential process to use. Following this sequence, you can obtain the other examples in the chapter.

MODE/FORMT: Change MODE to DifEq and set format by pressing GRAPH MORE FORMT FldOff. (Drawing slope field lines here would detract from our particular solution.)

F1: Define the second-order equation as a set of two first-order equations. The selected functions may or may not be graphed, depending on the AXES setting.

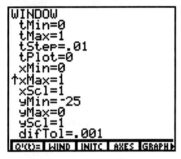

F2: Set the window. Since we know this solution is $y = -16x^2$, we set t to act like x and view negative y-values. The difTol setting indicates the RK setting, otherwise EStep (Euler step) would show.

➤ *Tip: The ZOOM options are rarely useful for differential equations because they are centered at the origin.*

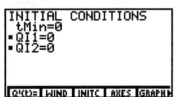

F3: Set initial conditions. Having specified FldOff, we must supply initial conditions. They are both zero in this case.

F4: Select axes. We want the graph of the particular solution, so we exclusively set y=Q1. Because we have selected both equations in the Q'(t)= screen, they would both be graphed if we used the inclusive setting y=Q.

F5: Graph. This graph is what we expected. At this point you can use DRAW, TRACE or EXPLR to investigate further.

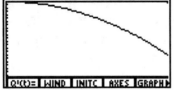

The second-order equation $s'' + \omega^2 s = 0$

Next we look at a second-order differential equation that depends on only one other variable. In this case s.

$$\frac{d^2 s}{dt^2} = F(s) = -\omega^2 s$$

This example of a differential equation describes simple harmonic motion, and it is known that the solution to this equation is

$$s(t) = C_1 \cos \omega t + C_2 \sin \omega t$$

For our example we will let $\omega = 4$ and then set the initial conditions to be $s(0) = 1$ and $s'(0) = -6$. Substituting these initial conditions into the general solution, we find the particular solution has constants $C_1 = 1$ and $C_2 = -3$.

Let's graph an approximation and compare it to this known solution. Since this is a trigonometric solution, we graph with $0 \leq x \leq 2\pi$.

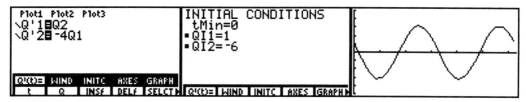

Figure 24.2 The graph of the solution to $s'' = -4s$ with window $0 \leq t \leq 2\pi$, $0 \leq x \leq 2\pi$, and $-5 \leq y \leq 5$.

The GRAPH MORE DRAW DrawF

We now make a common use of the DRAW menu. Once you have the particular solution graphed, you might want to compare it to an explicit analytical formula that you have derived. For example, in this case we want to graph $y = \cos(2x) - 3\sin(2x)$ and see if the graphs match. Pressing DRAW MORE DrawF (F2) displays the home screen with DrawF pasted and waiting to be completed with the function. In Figure 24.3 we see that the differences are hardly noticeable.

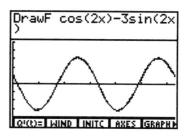

Figure 24.3 Use DrawF to graph a solution and compare it to the original graph.

The linear second-order equation $y'' + by' + cy = 0$

Equations of this type are called linear second-order equations because, when we isolate the second derivative, it is equal to a linear equation with y and y' as the variables and b and c as the coefficients. One application of this general equation is describing the motion of a spring.

The characteristic equation

The solution of a second-order linear equation hinges on the zeros of the quadratic equation $r^2 + br + c = 0$, namely

$$r = -\tfrac{1}{2}b \pm \tfrac{1}{2}\sqrt{b^2 - 4c}$$

There are three cases, based on the sign of the discriminant $b^2 - 4c$. You can use **POLY** to find the zeros, but it is easier to just write the solution in the above form. With the zeros of the equation and the initial values, you can use **SIMULT** to solve simultaneous equations for the constants C_1 and C_2 which appear in each of the three general solution equations.

The overdamped case: $y'' + by' + cy = 0$ with $b^2 - 4c > 0$

The general solution to this case is

$$y(t) = C_1 e^{r_1 t} + C_2 e^{r_2 t}$$

where r_1 and r_2 are zeros of the characteristic equation.

For example, suppose $b = 3$ and $c = 2$, then the discriminant $3^2 - 4(2) > 0$. Figure 24.4 shows the setup for graphing a particular solution with initial conditions $y = -.5$ and $y' = 3$ at $t = 0$. The window is $0 \le t \le 2\pi$, $0 \le x \le 2\pi$, and $-.6 \le y \le .6$. Thinking of this as a spring's motion in oil, we see that the spring swings past equilibrium ($y = 0$) and then its motion is damped until it is at rest. It never swings past the equilibrium after the first time.

Although this is not shown, you can check this graphical solution against the symbolic solution by using

$$\text{DrawF } 2e^{\wedge}(-X) - 2.5e^{\wedge}(-2X)$$

This solution ($C_1 = 2$ and $C_2 = -2.5$) was found by solving for C_1 and C_2 in simultaneous equations created from the initial conditions. See page 44.

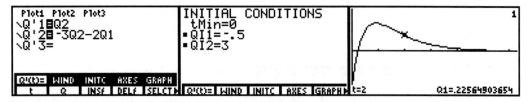

Figure 24.4 The setup and graph of the overdamped case with $0 \le x \le 2\pi$ and $-.6 \le y \le .6$.

The critically damped case: $y'' + by' + cy = 0$ with $b^2 - 4c = 0$

The general solution to this case is

$$y(t) = (C_1 t + C_2)e^{-bt/2}$$

where $b = -2r$ and r is the only zero of the characteristic equation.

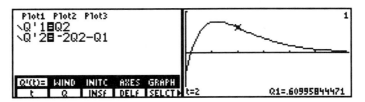

Figure 24.5 The setup and graph of the critically damped case.

For example, suppose $b = 2$ and $c = 1$, then the discriminant $2^2 - 4(1) = 0$. Figure 24.5 shows the particular solution with the same initial conditions $y = -.5$ and $y' = 3$ at $t = 0$, as in the last example. The window is $-1 \leq y \leq 1$. We see a graph that is similar to the solution of the previous case, but notice that it is not damped as quickly. Again using the initial values, you could find C_1 and C_2 to graph the solution

```
DrawF (2.5X-0.5)e^(-X)
```

The underdamped case: $y'' + by' + cy = 0$ with $b^2 - 4c < 0$

The general solution to this case is

$$y(t) = C_1 e^{\alpha t} \cos \beta t + C_2 e^{\alpha t} \sin \beta t$$

where $r = \alpha \pm i\beta$ are complex zeros of the characteristic equation.

For example, suppose $b = 2$ and $c = 2$, then the discriminant $2^2 - 4(2) < 0$. Figure 24.6 shows a home screen setup for the particular solution with initial conditions $y = 2$ and $y' = 0$ at $t = 0$. To enter function definitions easily, a **CUSTOM** screen was used to enter the prime mark ('). The initial conditions are variables and must be entered using the **STO→** command, not an equal sign. We graph the solution from $-.5 \leq y \leq 2$ and see that the motion is damped but crosses the equilibrium point at least twice in that interval. From the solution equation, we know that

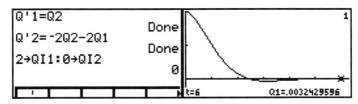

Figure 24.6 The setup and graph of the underdamped case.

this graph is an exponentially damped sine curve.

By using

```
DrawF 2e^(-X)(cos(X)+sin(x))
```

you can see how our graph compares to the graph of the actual solution.

25. THE LOGISTIC POPULATION MODEL

In this chapter we will derive a logistic equation to model the population growth of the United States. This is a difficult chapter because it will not have a detailed explanation of the theory behind the techniques that will be applied. It is suggested that the reader refer to the section "Models of Population Growth" in the differential equations chapter from *Calculus* by Hughes-Hallett, et al., (first or second edition). This chapter will parallel that exposition and show how the calculations and graphs can be shown using a TI-86 or TI-85. But in the end, this might be frustrating to some as we show how the logistic equation, which we struggled to derive, can be found by using a single command on the TI-86.

Entering U.S. population data 1790 - 1940

The first step is to enter the known data from the U.S. Census Bureau. Typically, annual data is not indexed by the year itself, but by years from a base year. Also, large numbers are usually rounded. Our base year is 1790 and we have rounded populations to the nearest tenth of a million. For example, the data pair for 1800 is (10, 5.3). We first enter the data as shown in Figure 25.1 You may prefer to enter the LY using the `seq()` command from the home screen as follows:

`seq(10*I,I,0,15)→LY`

Next use `2nd_STAT EDIT` to enter the data into LP as shown in Figure 25.1.

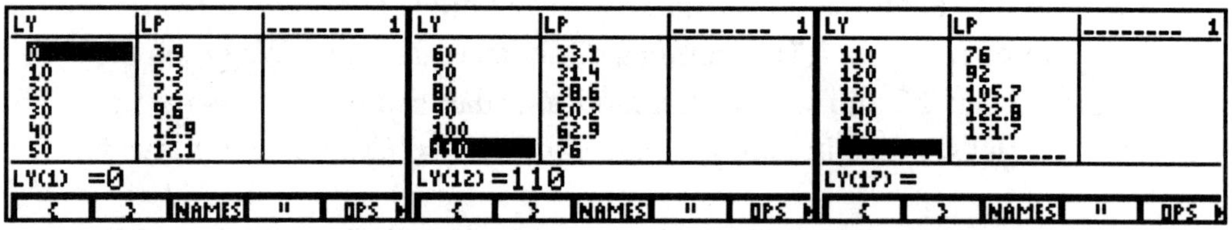

Figure 25.1 U.S. Census Data 1790 - 1940.

Estimating the relative growth rates: *P' / P*

We want to find the relative annual growth rate, but we settle for an approximation that uses data from the previous and next decade:

$$\frac{1}{P} \cdot \frac{dP}{dt} \approx \frac{1}{P_i} \cdot \frac{P_{i+1} - P_{i-1}}{20}$$

25. THE LOGISTIC POPULATION MODEL   147

Figure 25.2 shows the seq() command that will do all these calculations and store them in LRG. Beware that we had to start at I = 2 because we needed a previous decade for our calculations. Looking at the column of LRG values, we see that a rough estimate for the relative growth rate is 3%.

Figure 25.2 Using the seq() command to estimate the relative growth rate and storing it in LRG.

Modeling population growth with a simple exponential function

A naïve model builder would say the population starts at 3.9 (million) and grows at a 3% rate so that the following simple exponential model could be used:

$$y = 3.9e^{0.03x}$$

In Figure 25.3 we see the folly of this approach as the population for later years did not rise as fast as predicted. The graph of the actual data is shown as a decade step function given by y2. We have seen this int() trick before; here the y1 function is in years and the list is in decades, so we divide x by 10 so that the units match in y2.

The relative growth rate declines

By looking at the estimated relative growth rates for the years from 1860 to 1930, we see that the growth rate slows. We now want like to quantify this decline by

Figure 25.3 A simple model y1 becomes increasingly inaccurate for later years.

comparing $\frac{1}{P} \cdot \frac{dP}{dt}$ to P. This requires some data rearrangement. The estimated relative growth rate is in LRG but, because of the way it was calculated, it has no entry for the first or last decade (it has only 14 terms while there are 16 population entries). We make a new list with the corresponding populations to the LRG list by loading a copy of LP into LPX and then deleting the first and last entries. (It would not do to delete entries from LP since the calculations for LRG use all of the LP data.) To copy a list, move the cursor to highlight the empty heading and press ENTER, then type the new name LPX and press ENTER.

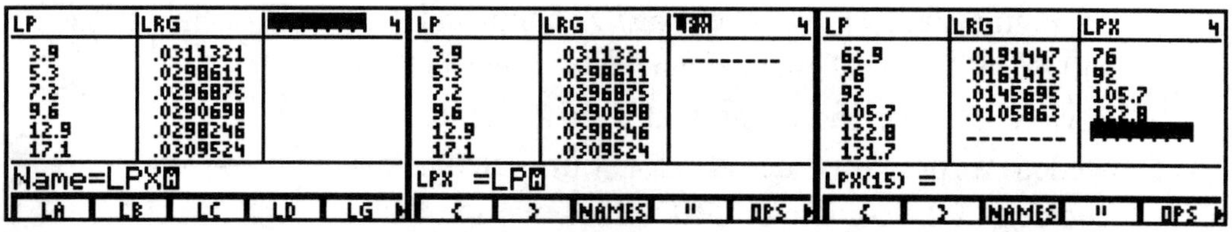

148 PART V / DIFFERENTIAL EQUATIONS

Figure 25.4 Copy **LP** *to* **LPX** *and delete the first and last entries of* **LPX** *so it matches entries in* **LRG**.

Now press **ENTER** and type the list name to be copied. (You could even specify a calculation on the list.) In Figure 25.4 we copy **LP** to **LPX** and use **DEL** to delete the first and last entries of **LPX**. The relative growth rates in **LRG** now correspond to the populations in **LPX**.

A scatterplot of *P' / P* against *P* with 2nd_STAT PLOT

One of the remarkable features of a TI calculator is its statistical capability. The fuller features can be explored using the *TI Guidebook*. We create a scatterplot to look for a relationship between the relative growth rate data, in **LRG**, and the corresponding population data, in **LPX**.

The **2nd_STAT PLOT** key will display a **STAT PLOTS** screen showing the setup status of the three stat plots. After selecting one, an individual setup window will allow you to set the following options:

> **ON OFF** same effect as the = and ▪ toggle for y(x)= functions
>
> **Type=** the first icon indicates scatterplot
>
> **Xlist Name=, Ylist Name=** list name for each axis
>
> **Mark=** graphic symbol for the plot points

▶ *Tip: The status of the three plots as on or off is also shown at the top of the* y(x)= *edit screen: plots that are on are highlighted. The status can be toggled in this screen by highlighting the plot name and pressing* **ENTER**.

▶ *Tip: A frequent annoyance is that a* y(x)= *function is left on from previous use and shows up on the graph screen. Even if it is outside the window, it still slows the graphing. Check the* y(x)= *screen for proper selections before graphing. Use* **MORE ALL-** *to quickly deselect all functions. (TI-85 users must check both the* y(x)= *and STAT PLOT screens.)*

In Figure 25.5, `PLOT1` has been selected, turned on, and selected as a scatterplot for `LPX` and `LRG`, with a box marker. Then we set an appropriate window. Pressing `GRAPH` will produce the graph shown.

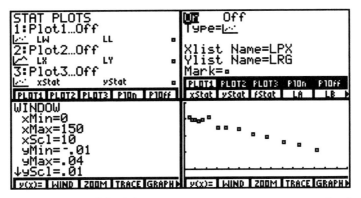

In Figure 25.5 `Plot1` *selected and graphed as a scatterplot for* `LRG` *against* `LPX` *with a box marker.*

▶ *Tip: The* `ZOOM ZDATA` *option sets a window that fits your data if you are unsure about the window settings.*

Finding a regression line to fit the data: `2nd_STAT CALC LinR`

Another statistical feature of the TI calculator is its ability to find regression equations of various kinds. The word regression is a statistics term, but for our purposes we will think of it as meaning a *best fit* type of equation. When we say regression line equation, we mean the equation of the line that comes closest to fitting our data. We also have exponential regression, the exponential equation that best fits the data. Our scatterplot in Figure 25.5 has a linear look, so we fit it using a linear regression equation.

Figure 25.6 shows the use of `2nd_STAT CALC LinR` (F3) with `LPX` and `LRG` specified to find the equation of the line $y \approx -.00017x + .0318$. By putting this equation in `y3` and graphing (with `Plot1` on), we see that the regression line fits this relationship.

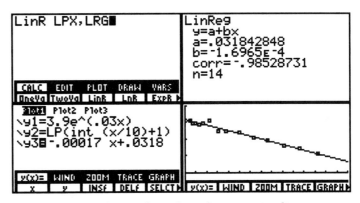

Figure 25.6 Finding and graphing the regression line $y \approx -0.00017x + 0.0318$.

► *Tip: The regression commands allow the optional addition of a function name where the regression equation (in full accuracy) is to be placed. We could have used* LinR LPX,LRG,y3 *in the previous example.*

Using the regression line to rewrite the differential equation

Now we transform the linear regression equation $y = ax + b$ back to its differential equation form with $y = \frac{1}{P} \cdot \frac{dP}{dt}$ and $x = P$. Next we multiply by P to get the differential equation

$$\frac{dP}{dt} = 0.0318P - 0.00017P^2$$

To see the nature of the solutions to this, we go to DifE◄ mode and then turn on the SlpFld option in GRAPH FORMT (as introduced in Chapter 22). In Figure 25.7 we see the setup process using the five GRAPH menu keys in their order.

This produces a graph of the particular solution with initial value of 3.9 for P. We see in Figure 25.7 that the graph rises over the entire interval and that the concavity changes from up to down. Also, there seems to be an upper limit on y-axis values. This is the general shape of a logistic curve.

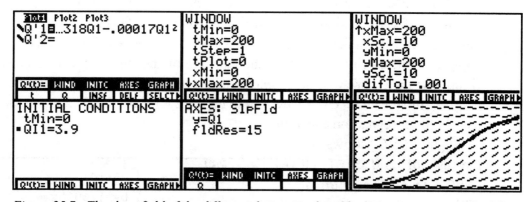

Figure 25.7. The slope field of the differential equation found by linear regression. The full equation is Q'1=.0318Q1-.00017Q1².

The general logistic equation

We are now ready for the traditional expression of a logistic equation,

$$P = \frac{L}{1 + Ae^{-kt}}$$

25. THE LOGISTIC POPULATION MODEL   151

We have seen that there appears to be an upper bound on the population. This makes sense from a biological point of view; the term 'carrying capacity' is often used to describe this limiting value.

Graphing the derived logistics equation

The carrying capacity is approximated by $L = b/-a$ (from the regression line $y = ax + b$), so in our population model example, $L \approx 187$. The variable k in the general formula is b. The parameter A is given by $A = (L - P_0)/P_0$; we have $P_0 = 3.9$, so $A \approx 47$. We can now write and graph the solution in its traditional form. We have found that

$$P = \frac{187}{1 + 47e^{-0.0318\,t}}$$

is the particular solution to

$$\frac{dP}{dt} = 0.0318P - 0.00017P^2.$$

The calculations for L and A are shown in Figure 25.8. We clear the initial condition for Q'1 so that we can see the graph of the defined function which graphs directly from the home screen. Notice that it matched the solution in Figure 25.7.

You may be concerned with the unrealistic value of L. It was derived from data before the baby boom, thus our model projects lower populations than we are experiencing. The current U.S. population is over 250 million, well beyond the value $L = 187$.

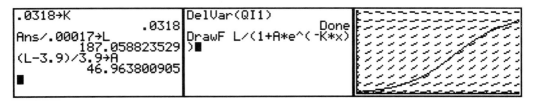

Figure 25.8 The parameters and graph of the traditional logistic equation.

The TI-86 built-in logistic regression feature: LgstR

It is usually worth the effort to go through and understand all the details once, but for practical use we welcome a more efficient method. The TI-86 has a simple command to model logistic data. We could have just entered the data in LY and LP then simply used the STAT CALC MORE LgstR command to find a logistics equation. In Figure 25.9 you will notice that there is an additional parameter d in the formula definition which causes a vertical shift. Also note that the negative sign in the exponent is incorporated into the parameter c. All values can be viewed by first scrolling down and then to the right. The parameter values are slightly different but very close to the ones we derived.

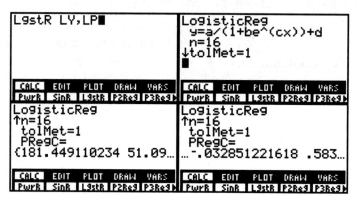

Figure 25.9 The logistic parameters calculated directly from the original lists.

➤ *Tip: Using the proper regression equation to model a situation is important. The various STAT CALC regression options allow you to try out different models on the same data to look for a good equation.*

26. SYSTEMS OF EQUATIONS AND THE PHASE PLANE

In this chapter we look at examples of systems of differential equations where the independent variable is time. Any two of these equations can be graphed to show their relationship to each other. Points on the graph are called the phase trajectory or orbit in the phase plane. Two popular examples of using systems of differential equations are the S-I-R model the predator-prey model.

For example, in a predator-prey model we can graph the predator population with time on the *x*-axis and then also graph the prey population with time on the *x*-axis. We can also create a phase plane graph with predator population on the *y*-axis and prey population on the *x*-axis.

As in the previous chapter, our goal here is to use the calculator's graphics to supplement detailed examples from *Calculus*, Hughes-Hallet, et al. Refer to that text for further background on the derivation and interpretation of the models.

The S-I-R model

The S-I-R stands for Susceptible - Infected - Recovered, so you can tell that this is used to model epidemics. The population is divided into the three groups and people move from S to I to R, or they just stay in S. The three rates in terms of time are

$$\frac{dR}{dt} = bI$$ the recovery rate depends upon the number infected

$$\frac{dS}{dt} = -aSI$$ the susceptible rate is negative and depends on both the number of infected and susceptible

$$\frac{dI}{dt} = aSI - bI$$ the infected rate is the negative of the sum of the other two rates.

Because knowing any two of the quantities *S*, *I*, *R* will automatically determine the third, we will concentrate on the last two rates.

The boarding school epidemic

The following example can be found in fuller detail in *Calculus*, Hughes-Hallet, et al. (second edition p. 563, first p. 542). There were 762 students in a boarding school and one returned from vacation with the flu. Two more students became sick the second day. This means we can approximate $a = -0.0026$ from

$$\frac{dS}{dt} = -aSI, \text{ since } (-2) = a(762)(1)$$

This flu lasts for a day or two, so we will assume half of the sick get well each day (thus $b = -0.5$).

Time plots for the model

In Figure 26.1 we define the differential equations and graph the variables S (Q1) and I (Q2) over time. This means we want two graphs (and no slope field), so we use **FORMT** to select **FldOff**. We look at the epidemic for 20 days, which means we set the window with **xMax=20** and the t-values are the same as x. There are just under 800 students, so we set **yMax=800**. Next we set the initial conditions then set the axes to **Q** so that both **Q1** and **Q2** will be graphed. We see that the susceptible are in a constant decline and that the number of infected peaks in the sixth day.

Phase plots for the model

In Figure 26.2 we look at the *SI* phase plane. The same definitions of **Q'1** and **Q'2** and initial conditions remain from the previous plot, but we now change the **AXES** to **x=Q1** and **y=Q2**. The model requires $0 \leq Q1 \leq 800$ and $0 \leq Q2 \leq 400$ and the time, $0 \leq t \leq 20$ days. In order to get a more accurate graph, we use **tStep=.1**. We also use **yMin=-100** to leave room at the bottom of the screen for trace values. Watching the screen as it graphs, you will notice that the graph is drawn from right to left, since the first point is $(S, I) = (762, 1)$. Using **TRACE** on this phase plot feels backward because the right arrow key will increase t which causes the trace cursor to move left. If you press the left arrow key at the start, then nothing will happen. We can see the peak of this curve has a horizontal axis value $Q1 \approx 200$; this we call the threshold value. You can use **TRACE** to find the peak more accurately at $(192, 306)$, where $t = 5.3$ days. The threshold value is of interest when we look at the slope field.

Figure 26.1 The graphs of S and I over time.

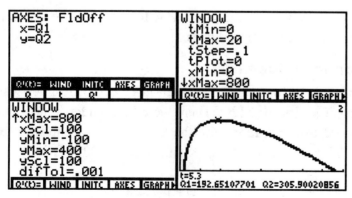

Figure 26.2 The SI phase plane.

➤ *Tip: The TRACE cursor moves through the points connected to the t values when the right arrow is pressed; the cursor movement is not necessarily from left to right.*

Direction fields for the S-I-R model

Like the slope fields for functions in terms of *t*, we can graph slope lines for the relationship of *I* to *S*. Unlike a slope field which has *t* on the *x*-axis, a direction field shows a relationship between two quantities, like Q1 and Q2, in your system. Time is not shown on either axis, but each point does have a time value implicitly attached. This can be seen when you trace a phase plot; the cursor starts at tMin and jumps by tStep.

We use FORMT to set DirFld on the bottom row. We check the axes to see that they have the default setting and then graph. The resulting graph is shown in Figure 26.3. Of particular interest is that along any particular solution, the peak (threshold value) is at the same value (*S* = 192) on the horizontal *S*-axis. But if *S* is less than 192, then the *I*-values decrease immediately.

Figure 26.3 The SI phase plane with direction field.

Predator-prey model

In this kind of model, we let *x* be the number of predators and *y* the number of prey. The details for this kind of system are found in the *TI-86 Guidebook*, pp. 258-259 (pp. 17-22 to 17-23 in the *TI-85 Guidebook*). In our example, we simplify greatly, setting all constants to 1. (This follows the approach taken in *Calculus*, Hughes-Hallet, et al., second edition p. 567, first p. 546.)

The simplified predator-prey system is

$$\frac{dy}{dt} = y - yx \text{ and } \frac{dx}{dt} = -x + yx$$

We will define the predator population *x* as Q2 and the prey population *y* as Q1 (with units in millions.) We will follow the presentation order of the previous model, first graphing the two populations over time.

Time plots for the predator-prey model

In Figure 26.4 the individual predator and prey differential equations with respect to time are entered and the five steps are shown (some with partial screens). Although initially you may have no idea how to set the window, a little trial-and-error can lead to the window shown in Figure 26.4. We made it wide enough to show the periodic nature of the two graphs. Not shown are the settings yMin=0 and yMax=3.

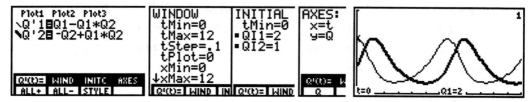

Figure 26.4 Time plots with $0 \le Q \le 3$. (Some partial screens shown.)

Phase plots for the predator-prey model

We now change the AXES and window to match the size of the predator and prey numbers. From the previous graph, we know that more than a complete cycle will take place if tMax=12, so we leave that setting but set xMax=3 since both populations are less than 3 (million). Remember when tracing values to press the right arrow key even though the trace cursor will at first move left; you are moving up in *t* which may correspond to moving right or left in *x*.

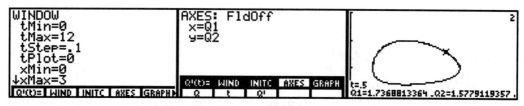

Figure 26.5 The periodic nature of the predator and prey populations.

➤ *Tip: When graphing a phase plot, the curve will have the style specified by the function defined on the x-axis.*

Direction field for the predator-prey model

By changing the GRAPH FORMT screen setting from FldOff to DirFld, we can superimpose the direction field over the phase plot. What we see from this direction field is that there is an equilibrium point at (1,1). At these values, the populations will theoretically remain stable and not have the kinds of cycles that we saw in Figures 26.5 and 26.6.

26. SYSTEMS OF EQUATIONS AND THE PHASE PLANE

The direction field of this system of two differential equations might look familiar; it was part of the last example in Chapter 22, but as the slope field for

$$\frac{dy}{dx} = \frac{-y + xy}{x - xy}$$

Using the chain rule, our system

$$\frac{dy}{dt} = y - yx \text{ and } \frac{dx}{dt} = -x + yx$$

gives that equation.

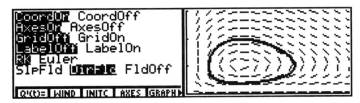

Figure 26.6 *The direction field over the phase plot.*

APPENDIX

Complex number form

Polar coordinates in the complex plane

Parametric graphing

Internet address information

Linking calculators

Linking to a computer

Troubleshooting

APPENDIX

Complex number form

The TI-86 and TI-85 accommodates complex numbers with ordered pairs where (a,b) represent $a + bi$, and $i = \sqrt{-1}$. However, the SOLVER will not accept complex numbers and finds only real solutions. The real and imaginary part answers to complex number calculations often extend off the screen. You can, of course, scroll over to see the whole number, but it is suggested that you either use MATH MISC MORE ▶Frac (F1) in hopes of a fraction or reset the Float setting in MODE to show a smaller number of digits (2, 3, or 4 are good settings). See Figures A.1 and A.2 for examples of these two techniques. The 2nd_CPLX menu has a list of operations that can be used on complex numbers.

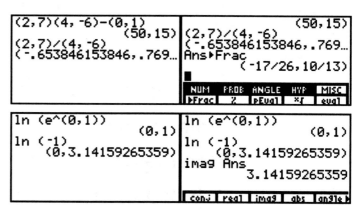

Figure A.1 Complex arithmetic examples and using the 2nd_CPLX menu.

Polar coordinates in the complex plane

The polar (Pol) graphing mode is based on complex numbers and is available on all TI graphing calculators.

Coordinate conversion

Each point in the Cartesian rectangular coordinate system has a polar coordinate form (r,θ) where r is the distance to the origin and θ is a measure of rotation from the x-axis. The TI representation is $(r\angle\theta)$ The related polar form is $re^{i\theta}$.

Figure A.2 Conversion techniques for complex numbers. Float 2 makes the answers more readable.

In Figure A.2 we see the coordinate conversion tools (▶Rec and ▶Pol) as found on the 2nd_CPLX MORE menu. The ▶Pol command will put a rectangular coordinate complex number in polar form. After the first calculation, we use MODE to make the output numbers rounded to two decimals, in the next calculations.

▶ *Tip: You will get an error message if you try to convert a real number to a complex one.*

▶ *Tip: The angle symbol ∠ is the 2nd option above the comma key.*

Polar graphing

One of the important aspects of graphing in the polar form or parametric form (next section) is that the curve need not be a function; the graphs need not pass the vertical line test. Polar graphing is explained fully in the *TI-86 Guidebook*, Chapter 8, so we give a very brief presentation here. (*TI-85 Guidebook*, Chapter Chapter 5)

Figure A.3 Polar graph of a circle in a ZSTD window

After changing MODE from Func to Pol, the y(x)= screen shows ri= function definitions. In Figure A.3, a circle of radius 5 is drawn with a ZSTD window. (The window ratio is not square, so the graph looks elliptical.)

In the polar graphing mode, the function definition key is now r(θ)= and equations are defined and graphed with θ as the independent variable. In Figure A.4, the equation $r = \theta$ is graphed. It is

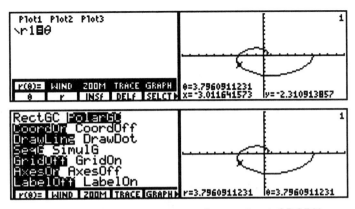

Figure A.4 The polar equation r=θ traced with GRAPH FORMT RectGC and then with PolarGC.

first traced with the default format setting (`RectGC`) and then changed to `PolarGC` by pressing GRAPH MORE FORMT. Notice in both settings that tracing by pressing the right arrow key increases θ values which may correspond to moving left or right in *x*.

Parametric graphing

The second option for graphing is `Par`, which stands for parametric. In this kind of graph, *x* and *y* are independently defined in terms of a variable *t* (usually thought of as time). The equation for a circle of radius 5 was easy in polar coordinates (Figure A.4). As a parametrically defined curve, a circle is composed of the sine and cosine functions. In Figure A.5 we set MODE to `Par`, define the equation, and set a standard window. Notice the changes in the y(x)= screen: equations are now in pairs and the independent variable is t. Also, TRACE now effects values of *t*, which satisfy $0 \leq t \leq 2\pi$ in the standard window.

Figure A.5 Set **MODE Par**, *define a pair of parametric equations, and graph in a standard window.*

Changing parameters

In Figure A.6 we look at the effect of changing the parameters of the equations defined in Figure A.5. The first curve appears the same. But by tracing you will see that the effect of increasing the coefficient of t to 3: the curve is wrapped around the same path three times. You could also use TABLE to see this difference. In the other

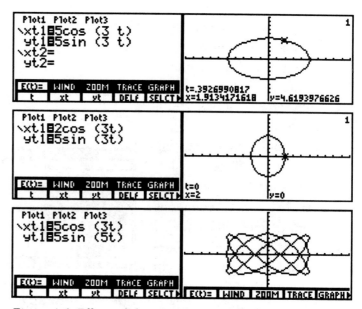

Figure A.6 Effects of changing the parameters.

screens, a lower coefficient on the cosine creates an ellipse and a mixture of periods results in a beautiful pattern.

► *Tip: The only ZOOM feature that will change the t-settings is ZSTD. All others ZOOM options will reset the window size, but not the t-values.*

Writing y=f(x) functions in parametric form

Any function $y = f(x)$ can be written in parametric form by setting xt1=t and yt1 to $f(t)$. We see in Figure A.7 that our first graph is restricted to the sine curve with $0 \le t \le 2\pi$ (The standard setting). In the next graph we have changed t so that the graph extends across the window.

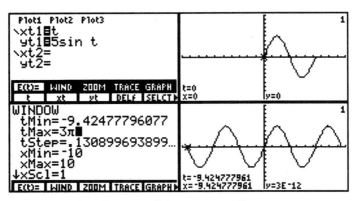

Figure A.7 Graphing y = f(x) parametrically.

Using logic to parametrically walk around the block

Just for fun, we conclude with some logically restricted definitions in a parametric definition that allow us to walk (trace, actually) around the block. (See the section on logical expressions in Chapter 22.) In the definitions, we multiply by logical expressions to restrict the definitions for each of the four sides of the unit square. We first use ZSTD and then ZIN to obtain the figure shown in Figure A.8. The figure can be shown as a square by using ZSQR.

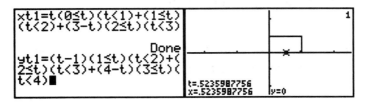

Figure A.8 A parametric equation draws a unit square.

Internet address information

The main internet address for Texas Instruments is

http://www.ti.com/

The calculator materials main menu can be obtained at

http://www.ti.com/calc/docs/

It is unlikely the above two entry addresses will change. Web page designs do change, but the following addresses are current as of January 1998 and will probably remain stable for finding the following topics.

Topic:	Address
Calculator Based Laboratory (CBL):	*http://www.ti.com/calc/docs/cbl.htm*
Calculator comparison:	*http://www.ti.com/calc/docs/gmtrx.htm*
Calculator support:	*http://www.ti.com/calc/docs/calcsupt.htm*
Classroom activities:	*http://www.ti.com/calc/docs/activities.htm*
Frequently Asked Question(FAQs):	*http://www.ti.com/calc/docs/faq.htm*
GraphLink:	*http://www.ti.com/calc/docs/link.htm*
Guides to downloading:	*http://www.ti.com/calc/docs/guides.htm*
Guidebook download:	*http://www.ti.com/calc/pdf/gb/83gb.pdf*
New calculator news:	*http://www.ti.com/calc/docs/calcnews.htm*
Program Archives:	*http://www.ti.com/calc/docs/arch.htm*
Ranger (like the CBL):	*http://www.ti.com/calc/docs/cbr.htm*
Resources:	*http://www.ti.com/calc/docs/resource.htm*

There are discussion groups available; you will find information about these from the main screen of http://www.ti.com/calc/docs/.

Linking calculators

The essentials of linking are presented in the *TI Guidebook* and will not be repeated here. But we include the following tips.

➤ *The end-jack must be pushed firmly into the socket. There is a final click you can feel as it makes the proper connection.*

➤ *If you are experiencing difficulty connecting, turn off both calculators, check the connection, and then turn them on and try again. If available, try other cables or calculators.*

➤ *When selecting items, the cursor square is hardly visible when the selection arrow is on the same item. As you arrow off of the item, double-check whether or not it has been selected.*

➤ *If you are required to drain your calculator memory before an exam, use* Back Up *to keep a copy on some other calculator. Even better is to store it on your computer, which is the next topic.*

Linking to a computer

The TI-Graph Link™ is a cable and software that connects a PC or Macintosh computer and your TI calculator. The software is available on the internet, so you can order just the cable. There are many advantages to using TI-Graph Link™:

- This is the best way to back up your work.
- It is the preferred way to write and edit programs.
- You can download and transfer programs from the internet archives.
- It allows you to capture the screen in a form for direct printing or use in a word processor.

Troubleshooting

Nothing shows on the screen

- Check the contrast.
- Check ON/OFF button.
- Pull out the batteries and correctly reinsert the batteries.
- As a last resort, remove the backup battery and reinsert all the batteries. Warning: This will erase all memory, including programs.

Nothing shows up on the graph screen except the axes

- Press TRACE to see if a function is defined but outside the window.
- There may be no functions selected.
- The function may be graphed along either axis and need the window reset.
- If there is a busy indicator (a running line) in the upper right corner of the screen, then the TI may be still calculating. Press ON if you can't wait.
- If there is a pause indicator (twinkling line) in the upper right corner of the screen, then the TI is paused from a program. Press ENTER to continue.
- If there is a full checkerboard cursor, then you have a full memory. You need to delete something; choose some things you no longer need or copy them to your computer.

Nothing shows up in the table

- You may have the Ask mode set and need to either enter x-values or change to Auto in the TBLSET.
- Check to see that a function is selected.
- Tables are not a feature on a TI-85.

I get a syntax error screen

The most common errors are

- Parenthesis mismatch. Count and match parentheses carefully.
- Subtraction vs. negative symbol. For example, the subtraction sign cannot be used to enter -10.
- Pasting a command in the wrong place. For example, a program name must be on a fresh line.
- Upper/lower case is wrong for variable names. For example, t and T are distinct variables. However, reserved words like sin and int can be mixed case and still recognized.

I get an UNDEFINED error message

- There may be an operation sign missing. For example, x*y is probably a values but xy would likely be undefined (the calculator thinks xy is the name of a single variable).
- Place a space or parenthesis to demarcate the function and variable. For example, sin1 is not the same as sin(1) or sin 1.

I get an error message

This can cover the widest array of problems. Read the message carefully; it will tip you off to the kind of error you are looking for. If you have no idea what could have caused it, consult the appendix of the *TI Guidebook* for explanations of the error messages.

I'm getting a result but it is wrong

Check for

- Using X in place of x. Use lower case x as the independent variable.
- Parenthesis mismatch. Count and match parentheses carefully.
- Subtraction vs. negative symbol. For example, the subtraction sign cannot be used to enter a -10.
- Order of operations. For example, 1/2x is (1/2)*x on a TI-86 but 1/2x is 1/(2*x) on a TI-85

- Correct default settings. For example, incorrect integrals may need to have `tol` reset.

My program won't run

Program errors are difficult to diagnose. Scroll and check your code with `PROGRAM EDIT`. Add temporary displays and pauses to check the progress and isolate the problem.

INDEX

2nd, 5
Acceleration, 63
ALPHA, 5
ALPHA-lock, 6
alpha-lock, 6
Anchorage annual rainfall, 106
Angles, 12
ANS, 8
Antiderivative, 86
Applications of the integral, 101
Approximating area, 80
Approximation accuracy, 54
Arc length, 101
Area approximating, 80
Asymptotic dangers, 31
Basic keys, 2
Boarding school epidemic, 153
Box with lid, 70
CATALOG, 14
Chain Rule, 65
Characteristic equation, 144
CLEAR, 7
Complex number form, 160
Composite functions, 17
Compounding, 104
Concavity, 62, 72
Convergence, 99
Coordinate conversion, 160
Correction keys, 7
Critically damped motion, 144
Cursor location, 27
CUSTOM menu, 15
Decimals to fractions, 9
Definite integral, 81
Degrees, 12
DEL, 7

Deltalst, 47, 60
der1, 53, 55, 71
Derivative
    at a point, 51, 52
    exact, 53
    graph, 56
    math definition, 58
    numeric, 53
    second, 61
    tangent function, 66
    to find the minimum, 69
    zero, 71
Difference equation, 137
Differential equation, 128
Differentiation rules, 64
Direction field, 157
Discrete vs. continuous, 104, 129
Distance from velocity, 76
DrawDot, 32
DrawF, 143
DrawLine, 32
Drug dosage, 116
dy/dx, 52
EDIT, 47
Editing keys, 7
EE, 11
ENTRY, 7
ERROR 27 NO SIGN CHNG, 43
Error messages, 166
Errors, common, 43
Errors in programs, 92
Euler's method, 137
Exact derivative, 53
EXIT, 15
Extrema, 36
FMAX, 36

`FMIN`, 36
`FMIN`, 69
`fnInt(nDer(...)...)`, 90
`fnInt(...,x)`, 84
`fnInt`, 82, 88, 103
Folding paper, 10
`For()...End`, 130
Force, 101
Format of numbers, 9
`FORMT`, 32
Fourier approximation, 123
Fourier series, 121
`FOURIER`, 124
`▸Frac`, 9
Function, 13
   composite, 17
   defining by lists, 18
   derivative graph, 56
   derivative, 56
   evaluating at a point, 16
   from inside a table, 22
   graphing, 24
   inverse graph, 34
   lists of values, 19
   new from old, 17
   periodic, 121, 126
   piecewise defined, 122
   selecting/deselecting, 21
   tables, 19
   triangle wave, 122
   values from the graph, 28
   zero from a table, 22
Fundamental Theorem of Calculus, 86
Future value, 103
General vs. particular solutions, 128
`GRAPH`, 24
Graphs, see specific topics
Greatest equation ever written, 7
Improper integrals, 96
Inaccurate graph, 32
`Indpnt`, 20
Infinite integrand, 99
Infinite limit of integration, 96
Infinite series, 119
`INFLC`, 73
Inflection point, 73
`INS`, 7
$\int f(x)$, 81
Integral answers (trusting), 100
Integral from a graph, 81
Integral upper limit as variable, 97
Integrand infinite, 99
Integration rules, 82
Intercept, 37
Internet address, 163
Intersection, 37
Interval of convergence, 111
Inverse functions, 34
Investments analysis, 41
`ISECT`, 37
Ladder problem, 68
Learning curve, 129, 131
Left-hand sums, 76, 79
`LgstR`, 152
Limit concept, 46
Limit graphically, 48
Limit numerically, 49
Linear second-order equations, 143
Linking calculators, 164
Linking to a computer, 165
`LinR`, 149
Lists, 46, 76
Logical expression, 121, 163
Logistic curve, 62
Logistic equation general, 151
Logistic population, 146
Malthus model, 17
`MATH`, 35, 51
Maximum/Minimum, 69
Mean, 105
Menu use, 15
`MODE`, 11
Navigation keys, 7
`nDer(fnInt(...)...)`, 90
`nDer`, 54, 67
Negative values in sum, 80

Normal distributions, 105
ON key, 2
Optimization, 68
Overdamped motion, 144
Parameters of geometric series, 118
Parametric form, 163
Parametric graphing, 162
Periodic function, 121, 126
Phase plane, 153
Phase plots, 154, 156
Piecewise defined functions, 122
Piggy-bank vs. trust, 120
Plot style, 32
Polar coordinates, 160
Polar graphing, 161
POLY, 40
Population, 146
Power functions, 96
Predator-prey, 155
Present value, 103
Pressure on a trough, 102
Product Rule, 64
Programs
    about, 91
    FOURIER, 123
    RSUM, 93
    SLPFLD, 134
    TAYLOR, 108
    trouble, 166
QUIT, 15
Quotient Rule, 65
Radians, 12
Recall, 7
Regression line, 149
Relative growth rates, 146
Riemann sums, 91
Right-hand sums, 76, 79
Rise and crash function, 121
Root, see Zero
ROOT, 35
RSUM, 92
Rules of differentiation, 64
Rules of integration, 82

S-I-R model, 153
Savings account, 117
Scatterplot, 148
Scientific keypad, 3
Scientific notation, 11
Screen contrast, 2
Second derivative concavity, 72
Second-key (2nd), 5
Second-order differential equations, 141
Selecting/deselecting a function, 21
seq, 78
Sequences summing, 78, 117
Series
    alternating harmonic, 113
    convergence, 111
    fast converging, 114
    finite geometric, 116
    Fourier, 121
    geometric, 116
    harmonic, 112
    infinite, 119
    parameters, 118
    slow converging, 115
    Taylor, 108
SIMULT, 44
Simultaneous equations, 44
Slope fields, 128, 132
Slope line, 51
SOLVER, 39, 41, 71, 103
Solver, 42
Solving an equation, 39
Speeding ticket, 50
Square wave function, 121
Standard deviation, 105
STAT EDIT, 77
STAT, 47
Statistical features, 106
Stored values, 6
STO→, 6
Sum negative values, 80
sum(seq(...)), 116
sum(seq(...)), 79
Systems of equations, 44, 153

Table scrolling, 21
Table second derivative, 63
Table unreliable, 50
TABLE, 19
Tangent, 66
TANLN, 51
Taylor polynomials, 108
TBLSET, 20
TI-85 subsections, 2, 16, 19, 23, 87, 134
Time plots, 154, 156
TRACE, 28, 70
Triangle wave function, 122, 126
Troubleshooting, 165
Underdamped motion, 145
Warnings, 50, 54
What-if analysis, 103
WIND, 24
Window panning, 29
Window setting, 25, 30
xMax, 25
xMin, 25
xRes, 25, 67
xScl, 25
y(x)=, 13, 24
YICPT, 37
yMax, 25
yMin, 25
yScl, 25
ZDECM, 26
Zeros, 35, 71
ZFIT, 26
ZINT, 28
ZOOM options, 33
ZOOM, 25
ZSTD, 26
ZTRIG, 26

NOTES

NOTES

NOTES

NOTES

NOTES

NOTES

NOTES